Walter Pfeifer

The Lie Algebras *su(N)*
An Introduction

Birkhäuser
Basel · Boston · Berlin

Author:
Walter Pfeifer
Stapfenackerweg 9
CH-5034 Suhr
Switzerland

2000 Mathematics Subject Classification: 17-01; 17B45, 17B81, 70G65, 81R12, 81R99

A CIP catalogue record for this book is available from the
Library of Congress, Washington D.C., USA

Bibliografische Information Der Deutschen Bibliothek
Die Deutsche Bibliothek verzeichnet diese Publikation in der Deutschen Nationalbibliografie;
detaillierte bibliografische Daten sind im Internet über <http://dnb.ddb.de> abrufbar.

ISBN 3-7643-2418-X Birkhäuser Verlag, Basel – Boston – Berlin

Member of the BertelsmannSpringer Publishing Group
Cover design: Micha Lotrovsky, CH-4106 Therwil, Switzerland
Printed on acid-free paper produced from chlorine-free pulp. TCF ∞
Printed in Germany
ISBN 3-7643-2418-X

9 8 7 6 5 4 3 2 1

www.birkhauser.ch

Contents

Preface

Lie algebras are not only an interesting mathematical field but also efficient tools to analyse the properties of physical systems. Concrete applications comprise the formulation of symmetries of Hamiltonian systems, the description of atomic, molecular and nuclear spectra, the physics of elementary particles and many others.

In particular, $su(N)$ Lie algebras very frequently appear and "there is hardly any student of physics or mathematics who will never come across symbols like $su(2)$ and $su(3)$" (Fuchs, Schweigert, 1997, p. XV). For instance, the algebra $su(2)$ describes angular momenta, $su(3)$ is related to harmonic oscillator properties (Talmi, 1993, p. 621) or to rotation properties of systems (Talmi, 1993, p. 797; Pfeifer, 1998, p. 113), and $su(4)$ represents states of elementary particles in the quark model (Greiner, Müller, 1994, p. 367).

This book is mainly directed to undergraduate students of physics or to interested physicists. It is conceived to give directly a concrete idea of the $su(N)$ algebras and of their laws. The detailed developments, the numerous references to preceding places, the figures and many explicit calculations of matrices should enable the beginner to follow. Laws which are given without proof are marked clearly and mostly checked with numerical tests. Knowledge of basic linear algebra is a prerequisite. Many results are obtained, which hold generally for (simple) Lie algebras. Therefore, the text on hand can make the lead-in to this field easier. For deepened studies of the representation theory, the book by Fulton and Harris (1991) is recommended.

The structure of the contents is simple. First, Lie algebras are defined and the $su(N)$ algebras are introduced starting from antihermitian matrices. In Chapter 3, the $su(2)$ algebras, their multiplets and the direct product of the multiplets are investigated. The treatment of the $su(3)$ multiplets in Chapter 4 is more labour-intensive. Casimir operators and methods to determine the dimensions of multiplets are introduced. In Chapter 5, the $su(4)$ multiplets are represented three-dimensionally. Making use of the $su(4)$ algebra, the properties of the Cartan–Weyl basis are demonstrated. Chapter 6 points to general relations of the $su(N)$ algebras.

Any reader who detects errors is asked to contact the author via the email address mailbox@walterpfeifer.ch. Of course, the author is willing to answer questions.

Chapter 1

Lie algebras

In this chapter the Lie algebra is defined and the properties of its underlying vector space are described. Discussing the role of the basis elements of the algebra one is lead to the structure constants, to some of their symmetry properties and to their relationship to the adjoint matrices. With these matrices the Killing form is constructed. As a natural example for a Lie algebra, general square matrices are looked at. The notion of "simplicity" is introduced.

Operators which constitute a Lie algebra act on vector spaces of functions. By means of the corresponding expansion coefficients, properties of these operators are shown. The matrices of the expansion coefficients make up a representation of the algebra. If this representation is reducible it can be transformed to the equivalent block diagonal form. The functions which are assigned to an irreducible representation form a multiplet.

1.1 Definition and basic properties

1.1.1 What is a Lie algebra?

A Lie algebra $\mathcal{L}$ comprises the elements $a, b, c, \ldots$, which may be general matrices or matrices with certain properties (real/complex matrix elements, internal symmetries) or linear operators etc. The elements can be combined in two different ways. To come first, the elements of a Lie algebra must be able to form Lie products $[a, b]$, which are also named Lie brackets or commutators. For square matrices $\underline{\underline{a}}$ and $\underline{\underline{b}}$, the well-known relation

$$[\underline{\underline{a}}, \underline{\underline{b}}] = \underline{\underline{ab}} - \underline{\underline{ba}} \tag{1.1}$$

defines a commutator, which, of course, is again a matrix. On the other hand, if the elements of a Lie algebra are operators or more general quantities, the commutator $[a, b]$ has still to be defined, but the right hand side of (1.1) needs not to be satisfied. We want that, in addition to the formation of the commutator, it must be possible to combine the elements of the Lie algebra linearly, i.e., they constitute a vector space.

Thus, the **definition of a Lie algebra $\mathcal{L}$** demands the following properties of its elements $a, b, c, \ldots$.

a) the commutator of two elements is again an element of the algebra

$$[a, b] \in \mathcal{L} \qquad \text{for all } a, b \in \mathcal{L}, \tag{1.2}$$

b) a linear combination $\alpha a + \beta b$ of the elements a and b with the real (or complex) scalars α and β is again an element of the algebra, i.e.,

$$\alpha a + \beta b \in \mathcal{L}, \quad \text{if } a, b \in \mathcal{L}. \tag{1.3}$$

Therefore the element 0 (zero) belongs to the algebra.

c) The following linearity is postulated

$$[\alpha a + \beta b, c] = \alpha [a, c] + \beta [b, c] \qquad \text{for all } a, b, c \in \mathcal{L}. \tag{1.4}$$

d) Interchanging both elements of a commutator results in the relation

$$[a, b] = -[b, a]. \tag{1.5}$$

With (1.5) and (1.4) one proves that also $[a, \beta b + \gamma c] = \beta [a, b] + \gamma [a, c]$ holds. Of course, we have $[c, c] = 0$.

e) Finally, the Jacobi identity has to be satisfied as follows

$$[a, [b, c]] + [b, [c, a]] + [c, [a, b]] = 0. \tag{1.6}$$

Using (1.1) one shows that this identity holds for square matrices.

Note that we do not demand that the commutators are associative, i.e., the relation $[a, [b, c]] = [[a, b], c]$ is not true in general. Some authors choose other logical sequences of postulates for the Lie algebra. For instance one can define $[c, c] = 0$ and deduce (1.5) (Carter, 1995, p. 5).

f) In addition to a) up to e) we demand that a Lie algebra has a finite dimension n, i.e., it comprises a set of n linearly independent elements $e_1, e_2, \ldots, e_n$, which act as a basis, by which every element x of the algebra can be represented uniquely like this:

$$x = \sum_{j=1}^{n} \xi_j e_j. \tag{1.7}$$

In other words, the algebra constitutes an n-dimensional vector space. Sometimes, the dimension is named order. If the coefficients ξ_j in (1.7) and α, β in (1.3) are real, the algebra is named real. In a complex or complexified algebra the coefficients are complex.

We summarise points a) up to f): *A Lie algebra is a vector space with an alternate product satisfying the Jacobi condition.*

In accordance with the definition point e), the basis elements e_i meet the **Jacobi identity** (1.6). If this is the case, the arbitrary elements

$$x = \sum_{j=1}^{n} \xi_j e_j \qquad y = \sum_{i}^{n} \eta_i e_i \qquad \text{and} \qquad z = \sum_{k}^{n} \zeta_k e_k \tag{1.8}$$

satisfy the identity as well. We introduce a symbol for the Jacobi form

$$\{a, b, c\} \equiv [a, [b, c]] + [b, [c, a]] + [c, [a, b]] \,, \tag{1.9}$$

which is linear in a, b, c. Replacing these elements by x, y and z and inserting the expressions for x, y and z we obtain

$$\{x, y, z\} = \sum\nolimits_{jik}^{n} \xi_j \eta_i \zeta_k \{e_j, e_i, e_k\} \,. \tag{1.10}$$

Since every Jacobi form on the right hand side vanishes, it is proved that in order to have the identity (1.6), it suffices to ask that the condition is satisfied by the basis elements.

Clearly, the **choice of the basis** $e_1, e_2, \dots, e_n$ is arbitrary. With a nonsingular matrix $\underline{\underline{r}}$, which contains the real (or complex) numbers r_{kl}, a new basis $e'_1, e'_2, \dots, e'_n$ can be built this way:

$$e'_k = \sum_{l=1}^{n} r_{kl} e_l. \tag{1.11}$$

The new elements meet the Jacobi identity, which is shown in (1.10). It is a well-known fact that the elements e'_k are linearly independent if $e_1, e_2, \dots, e_n$ do and if $\underline{\underline{r}}$ is non-singular. Of course, a change of the basis of a real (or complex) Lie algebra by means of real (or complex) coefficients r_{kl} in (1.11) restores the algebra, and it keeps its name.

1.1.2 The structure constants

Due to (1.2) the commutator of two basis elements belongs also to the algebra and, following (1.7) it can be written like this

$$[e_i, e_k] = \sum_{l=1}^{n} C_{ikl} e_l. \tag{1.12}$$

The n^3 coefficients C_{ikl} are called **structure constants** relative to the $\{e_i\}$-basis. They are not invariant under a transformation as per (1.11) (see (1.17)). Given the set of basis elements, the structure constants specify the Lie algebra completely. Of course, a Lie algebra with complex structure constants is complex itself. Clearly, if the structure constants are chosen real, a real Lie algebra can be constructed on its basis.

The commutator of the elements x and y, (1.8), can be expressed by the basis elements as follows:

$$[x, y] = \sum\nolimits_{i,j}^{n} \xi_i \eta_j \, [e_i, e_j] = \sum\nolimits_{ijk} \xi_i \eta_j C_{ijk} e_k. \tag{1.13}$$

Analogously, the structure constants of the new basis $e'_1, e'_2, \ldots, e'_n$, (1.11), are determined this way:

$$[e'_i, e'_k] = \sum_{jl} r_{ij} r_{kl} \, [e_j, e_l] = \sum_{jlm} r_{ij} r_{kl} C_{jlm} e_m. \tag{1.14}$$

Since $\underline{\underline{r}}$ is nonsingular, the inverse matrix $\underline{\underline{r^{-1}}}$ exists and

$$\sum_p \left(r^{-1}\right)_{mp} e'_p = e_m \text{ holds,} \tag{1.15}$$

which we insert in (1.14) like this:

$$[e'_i, e'_k] = \sum_{jlmp} r_{ij} r_{kl} C_{jlm} \left(r^{-1}\right)_{mp} e'_p. \tag{1.16}$$

Thus, the structure constants of the new basis read

$$C'_{ikp} = \sum_{jlm} r_{ij} r_{kl} C_{jlm} \left(r^{-1}\right)_{mp}. \tag{1.17}$$

We insert (1.12) in the Jacobi identity (1.6) and obtain

$$\begin{aligned} 0 &= [e_i, [e_j, e_k]] + [e_j, [e_k, e_i]] + [e_k, [e_i, e_j]] \\ &= \sum_l C_{jkl} \, [e_i, e_l] + \sum_l C_{kil} \, [e_j, e_l] + \sum_l C_{ijl} \, [e_k, e_l] \\ &= \sum_{lm} C_{jkl} C_{ilm} e_m + \sum_{lm} C_{kil} C_{jlm} e_m + \sum_{lm} C_{ijl} C_{klm} e_m . \end{aligned} \tag{1.18}$$

Because the basis elements e_m are linearly independent, we get n equations for given values i, j, k like this:

$$0 = \sum_l \left(C_{jkl} C_{ilm} + C_{kil} C_{jlm} + C_{ijl} C_{klm}\right), \quad m = 1, \ldots, n. \tag{1.19}$$

We write the antisymmetry relation (1.5) of commutators as $[e_i, e_k] = -[e_k, e_i]$ and insert equation (1.12), which yields $\sum_l C_{ikl} e_l = -\sum_l C_{kil} e_l$. Due to the linear independence of the basis elements, we obtain n equations for given values i, k of the following form:

$$C_{ikl} = -C_{kil}, \quad l = 1, \ldots, n. \tag{1.20}$$

In Section 2.3 we will see that in $su(N)$-algebras the structure constants relative to the $\{e_i\}$-basis are antisymmetric in all indices and not only in the first two ones like in (1.20).

1.1.3 The adjoint matrices

Here we introduce the **adjoint** or **$\underline{\underline{ad}}$-matrices** . As per (1.2) the commutator of an arbitrary element a of a Lie algebra with a basis element e_i must be a linear combination of the basis elements similarly to (1.12). Therefore we write

$$[a, e_k] = \sum_{l=1}^{n} (ad(a))_{lk} e_l \qquad \text{for } i = 1, 2, \ldots, n. \tag{1.21}$$

The coefficients $(ad(a))_{lk}$ are elements of the $n \times n$-matrix $\underline{\underline{ad}}(a)$. It is easy to show that it is linear in the argument like this:

$$\underline{\underline{ad}}(\alpha a + \beta b) = \alpha \underline{\underline{ad}}(a) + \beta \underline{\underline{ad}}(b) \tag{1.22}$$

and that the following commutator relation holds:

$$\underline{\underline{ad}}([a, b]) = \left[\underline{\underline{ad}}(a), \underline{\underline{ad}}(b)\right]. \tag{1.23}$$

Replacing a by e_i in (1.21) yields

$$[e_i, e_k] = \sum_{l=1}^{n} (ad(e_i))_{lk} e_l. \tag{1.24}$$

Comparing with (1.12) we obtain

$$(ad(e_i))_{lk} = C_{ikl}. \tag{1.25}$$

We will come back to the adjoint matrices in Sections 1.4 and 3.1.

These $\underline{\underline{ad}}$ matrices appear also in the so-called **"Killing form"**, which plays an important part in the analysis of the structure of algebras.

1.1.4 The Killing form

The "Killing form" $B(a, b)$ corresponding to any two elements a and b of a Lie algebra is defined by

$$B(a, b) = Tr\left(\underline{\underline{ad}}(a) \cdot \underline{\underline{ad}}(b)\right). \tag{1.26}$$

The symbol Tr denotes the trace of the matrix product. It can be shown that the Killing form $B(a,b)$ is symmetric and bilinear in the elements a and b. We write the Killing form of the basis elements e_i and e_j of a Lie algebra

$$B(e_i,e_j) = Tr\left(\underline{\underline{ad}}(e_i)\,\underline{\underline{ad}}(e_j)\right) = \sum_l \left(\sum_k (ad(e_i))_{lk}\,(ad(e_j))_{kl}\right). \tag{1.27}$$

With (1.25) we have

$$B(e_i,e_j) = \sum_{l,k} C_{ikl}C_{jlk}. \tag{1.28}$$

In Sections 2.3 and 3.1 we will deal with the Killing form of $su(N)$ and $su(2)$, respectively.

1.1.5 Simplicity

From the theoretical point of view it is important to know whether a Lie algebra is **simple**. We give the appropriate definitions concisely.
A Lie algebra is said to be simple if it is not Abelian and does not possess a proper invariant Lie subalgebra.
The terms used here are defined as follows:

- A Lie algebra $\mathcal{L}$ is said to be "Abelian" if $[a,b]=0$ for all $a,b\in\mathcal{L}$. Thus, in an Abelian Lie algebra all the structure constants are zero.

- A *subalgebra* $\mathcal{L}'$ of a Lie algebra $\mathcal{L}$ is a subset of elements of $\mathcal{L}$ that themselves form a Lie algebra with the same commutator and field as that of $\mathcal{L}$. This implies that $\mathcal{L}'$ is real if $\mathcal{L}$ is real and $\mathcal{L}'$ is complex if $\mathcal{L}$ is complex.

- A subalgebra $\mathcal{L}'\neq\{0\}$ is said to be *proper* if at least one element of $\mathcal{L}$ is not contained in $\mathcal{L}'$.

- A subalgebra $\mathcal{L}'$ of a Lie algebra $\mathcal{L}$ is said to be *invariant* if $[a,b]\in\mathcal{L}'$ for all $a\in\mathcal{L}'$ and $b\in\mathcal{L}$. An invariant subalgebra is also named *ideal.*

We will meet simple Lie algebras in Sections 2.2 and 4.2.

1.1.6 Example

Obviously, **square$N$-dimensional matrices** ($N\times N$-matrices, sometimes called matrices of order N) constitute a Lie algebra. The commutators and linear combinations are again N-dimensional matrices and the conditions a) up to e) are satisfied. The basis matrices for the real space of complex matrices can be chosen like this, where i$=\sqrt{-1}$:

$$\begin{gathered}
\begin{pmatrix} 1 & 0 & & 0 \\ 0 & 0 & & \\ & & \ddots & \\ 0 & & & 0 \end{pmatrix}, \begin{pmatrix} 0 & 1 & & 0 \\ 0 & 0 & & \\ & & \ddots & \\ 0 & & & 0 \end{pmatrix}, \ldots, \\
\begin{pmatrix} 0 & 0 & & 0 \\ 1 & 0 & & \\ & & \ddots & \\ 0 & & & 0 \end{pmatrix}, \begin{pmatrix} 0 & 0 & & 0 \\ 0 & 1 & & \\ & & \ddots & \\ 0 & & & 0 \end{pmatrix}, \ldots, \begin{pmatrix} 0 & 0 & & 0 \\ 0 & 0 & & \\ & & \ddots & \\ 0 & & & 1 \end{pmatrix}, \\
\begin{pmatrix} \mathrm{i} & 0 & & 0 \\ 0 & 0 & & \\ & & \ddots & \\ 0 & & & 0 \end{pmatrix}, \begin{pmatrix} 0 & \mathrm{i} & & 0 \\ 0 & 0 & & \\ & & \ddots & \\ 0 & & & 0 \end{pmatrix}, \ldots, \\
\begin{pmatrix} 0 & 0 & & 0 \\ \mathrm{i} & 0 & & \\ & & \ddots & \\ 0 & & & 0 \end{pmatrix}, \begin{pmatrix} 0 & 0 & & 0 \\ 0 & \mathrm{i} & & \\ & & \ddots & \\ 0 & & & 0 \end{pmatrix}, \ldots, \begin{pmatrix} 0 & 0 & & 0 \\ 0 & 0 & & \\ & & \ddots & \\ 0 & & & \mathrm{i} \end{pmatrix}.
\end{gathered} \tag{1.29}$$

The set (1.29) can be written using the square matrices $\underline{\underline{E}}_{ij}$, which show the value 1 at the position (i,j) and zeros elsewhere. The canonical basis $\left\{\underline{\underline{E}}_{ij}\right\} \oplus \mathrm{i}\left\{\underline{\underline{E}}_{ij}\right\}$ forms the basis (1.29).

Using these $2N^2$ basis elements the definition point f) is met with real coefficients ξ_i in (1.7). In a word, the complex matrices of order N form a real vector space of dimension $2N^2$. On the other hand, the same vector space of complex $N \times N$-matrices can be constructed using complex coefficients and the basis $\left\{\underline{\underline{E}}_{ij}\right\}$. The resulting complex algebra is named $m(N,\mathbb{C})$ and it has the dimension N^2. In Section 2.2 we will interrelate it with $u(N)$.

1.2 Isomorphic Lie algebras

A Lie algebra with elements $a, b, \ldots$ is isomorphic to another Lie algebra with elements $A, B, \ldots$ if an unambiguous mapping $\phi(a) = A$, $\phi(b) = B$, ... exists which is reversible, i.e., $\phi^{-1}(A) = a$, $\phi^{-1}(B) = b$, ... It has to meet the relations

$$\begin{aligned} \phi(\alpha a + \beta b) &= \alpha\phi(a) + \beta\phi(b) \\ \phi([a,b]) &= [\phi(a), \phi(b)] \end{aligned} \tag{1.30}$$

for all scalars α and β.

The structure of both Lie algebras is identical and both have the same dimension n. Suppose $e_1, e_2, \ldots, e_n$ are basis elements of the first Lie algebra, then $\phi(e_1), \phi(e_2), \ldots, \phi(e_n)$ is a basis of the isomorphic Lie algebra. Of course, two isomorphic Lie algebras have the same structure constants.

Here the **Ado Theorem** is given without proof:
Every abstract Lie algebra is isomorphic to a Lie algebra of matrices with the commutator defined as in equation (1.1).

Consequently, the properties of Lie algebras can be studied by investigating the relatively vivid matrix algebras.

1.3 Operators and functions

1.3.1 The general set-up

In addition to the definition points a) up to f) of Section 1.1 we demand that operators which form a Lie algebra act linearly on functions. These functions or states or "kets" make up a d-dimensional vector space with linearly independent basis functions $\psi_1, \psi_2, \ldots, \psi_d$ or $|1\rangle, |2\rangle, \ldots, |d\rangle$. That is, an arbitrary function ψ of this space can be written as

$$\psi = \sum_{k=1}^{d} c_k \psi_k \tag{1.31}$$

with real or complex coefficients c_k .

We demand that the operators of the Lie algebra acting on a function produce a transformation of the function so that the resulting function lies still in the original vector space. If we let act the element $\boldsymbol{x}$ of the Lie algebra on the basis function ψ_k, it yields the following linear combination of basis functions

$$\boldsymbol{x}\,\psi_k = \sum_{l=1}^{d} \Gamma(\boldsymbol{x})_{lk} \psi_l. \tag{1.32}$$

Notice the sequence of the indices in Γ. Making use of (1.31) we can write down the action of $\boldsymbol{x}$ on an arbitrary function ψ like this:

$$\boldsymbol{x}\,\psi = \sum_{k,l=1}^{d} c_k \Gamma(\boldsymbol{x})_{lk} \psi_l, \tag{1.33}$$

which is a consequence of the linearity of $\boldsymbol{x}$. In Section 1.4 we will see that $\Gamma(\boldsymbol{x})_{lk}$ is linear in $\boldsymbol{x}$, i.e., $\Gamma(\boldsymbol{x})_{lk} = \Gamma\left(\sum_{i=1}^{n} \xi_i \boldsymbol{e}_i\right)_{lk} = \sum_{i=1}^{n} \xi_i \Gamma(\boldsymbol{e}_i)_{lk}$. Therefore, equation

(1.33) results in

$$\boldsymbol{x}\,\psi = \left(\sum_{i=1}^{n} \xi_i \boldsymbol{e}_i\right)\psi = \sum_{i=1}^{n}\sum_{k,l=1}^{d} \xi_i c_k \Gamma\left(\boldsymbol{e}_i\right)_{lk} \psi_l. \tag{1.34}$$

We see that the action of the operators constituting a Lie algebra on the functions is mainly described by the coefficients $\Gamma\left(\boldsymbol{e}_i\right)_{lk}$.

Furthermore, we demand that **the vector space of the functions is an inner product space**. That is to say, for every pair of functions ψ and φ, an inner product $\langle\psi \mid \varphi\rangle$ is defined with the following well-known properties:

i. $\langle\psi \mid \varphi\rangle = \langle\varphi \mid \psi\rangle^*$ (complex conjugate)
ii. $\langle\psi \mid \varphi + \varphi'\rangle = \langle\psi \mid \varphi\rangle + \langle\psi \mid \varphi'\rangle$
iii. $\langle\psi \mid c\varphi\rangle = c\,\langle\psi \mid \varphi\rangle$ (c: complex or real number)
iv. $\langle\psi \mid \psi\rangle \geq 0$
v. $\langle\psi \mid \psi\rangle = 0$ if and only if $\psi = 0$.

Note that we obtain $\langle c\psi \mid \varphi\rangle = c^*\,\langle\psi \mid \varphi\rangle$ from i and ii.

In an inner product vector space with linearly independent basis functions $\psi_1, \psi_2, \ldots, \psi_d$ it is always possible to construct an orthogonal basis $\psi_1^0, \psi_2^0, \ldots, \psi_d^0$, which satisfies

$$\langle\psi_i^0 \mid \psi_k^0\rangle = \delta_{ik} \quad i,k = 1,2,\ldots,d. \tag{1.35}$$

In the following, we will presuppose that the basis functions are orthonormalized, which is not a restriction. Assuming this property we form the inner product of both sides of eq. (1.32) with the basis function ψ_m:

$$\langle\psi_m \mid \boldsymbol{x}\psi_k\rangle = \sum_{l=1}^{d} \Gamma(\boldsymbol{x})_{lk}\,\langle\psi_m \mid \psi_l\rangle = \sum_{l=1}^{d} \Gamma(\boldsymbol{x})_{lk}\,\delta_{ml} = \Gamma(\boldsymbol{x})_{mk}\,. \tag{1.36}$$

In Section 1.4 we will refer to the square matrix $\underline{\underline{\Gamma}}(\boldsymbol{x})$, which contains the matrix elements $\Gamma(\boldsymbol{x})_{mk}$.

1.3.2 Further properties

We look into further properties of the operators constituting a Lie algebra and of the affiliated basic functions. In analogy to (1.36) we make up the inner product of both sides of eq. (1.31) with the basis function ψ_m as follows:

$$\langle\psi_m \mid \psi\rangle = \sum_{k=1}^{d} c_k\,\langle\psi_m \mid \psi_k\rangle = c_m, \tag{1.37}$$

which we insert again in eq. (1.31) like this:

$$\psi = \sum_{j=1}^{d} \psi_j \langle \psi_j \mid \psi \rangle . \tag{1.38}$$

The functions ψ and ψ_j are ket states and can be marked by the ket symbol this way:

$$|\psi\rangle = \sum_{j=1}^{d} |\psi_j\rangle \langle \psi_j \mid \psi \rangle . \tag{1.39}$$

Formally, we can regard the expression $\sum_{j=1}^{d} |\psi_j\rangle \langle \psi_j|$ as an operator which restores the state $|\psi\rangle$ like the identity operator $\mathbf{1}$. This is the *completeness relation*

$$\sum_{j=1}^{d} |\psi_j\rangle \langle \psi_j| = \mathbf{1} . \tag{1.40}$$

We apply it in order to formulate a **matrix element of a sequence of two operators**, say $\boldsymbol{x}$ and $\boldsymbol{y}$, like this:

$$\langle \psi_i \mid \boldsymbol{x}\boldsymbol{y}\psi_k \rangle = \langle \psi_i \mid \boldsymbol{x}\,\mathbf{1}\,\boldsymbol{y}\psi_k \rangle = \sum_{j=1}^{d} \langle \psi_i \mid \boldsymbol{x}\psi_j \rangle \langle \psi_j \mid \boldsymbol{y}\psi_k \rangle . \tag{1.41}$$

Making use of (1.36), we have the result

$$\Gamma(\boldsymbol{x}\boldsymbol{y})_{ik} = \sum_{j=1}^{d} \Gamma(\boldsymbol{x})_{ij}\, \Gamma(\boldsymbol{y})_{jk} \text{ or } \underline{\underline{\Gamma}}(\boldsymbol{x}\boldsymbol{y}) = \underline{\underline{\Gamma}}(\boldsymbol{x})\, \underline{\underline{\Gamma}}(\boldsymbol{y}) . \tag{1.42}$$

I.e., the matrices $\underline{\underline{\Gamma}}$ multiply in analogy with the corresponding operators.

Now, we investigate the matrix $\underline{\underline{\Gamma}}(\boldsymbol{x})$ containing the matrix elements $\Gamma(\boldsymbol{x})_{ik} = \langle \psi_i \mid \boldsymbol{x}\psi_k \rangle$, the adjoint of this matrix and adjoint operators. By definition, the adjoint matrix of $\underline{\underline{\Gamma}}(\boldsymbol{x})$, $\underline{\underline{\Gamma}}(\boldsymbol{x})^\dagger$, is transposed and conjugate complex, i.e.,

$$\Gamma(\boldsymbol{x})^\dagger_{ik} = \langle \psi_k \mid \boldsymbol{x}\psi_i \rangle^* . \tag{1.43}$$

Condition i of the inner product definition yields

$$\Gamma(\boldsymbol{x})^\dagger_{ik} = \langle \boldsymbol{x}\psi_i \mid \psi_k \rangle . \tag{1.44}$$

We define the adjoint operator $\boldsymbol{x}^\dagger$ by

$$\Gamma(\boldsymbol{x})^\dagger_{ik} = \langle \boldsymbol{x}\psi_i \mid \psi_k \rangle = \langle \psi_i \mid \boldsymbol{x}^\dagger \psi_k \rangle . \tag{1.45}$$

Using this relation, we write the matrix element of a sequence of two operators this way: $\langle \boldsymbol{xy}\psi_i \mid \psi_k\rangle = \langle \boldsymbol{y}\psi_i \mid \boldsymbol{x}^\dagger\psi_k\rangle = \langle \psi_i \mid \boldsymbol{y}^\dagger\boldsymbol{x}^\dagger\psi_k\rangle$ or $\langle \boldsymbol{xy}\psi_i \mid \psi_k\rangle = \left\langle \psi_i \,\middle|\, (\boldsymbol{xy})^\dagger\,\psi_k\right\rangle$. Thus we have found

$$(\boldsymbol{xy})^\dagger = \boldsymbol{y}^\dagger\boldsymbol{x}^\dagger \tag{1.46}$$

in accordance with the matrix relation (2.9).

1.4 Representation of a Lie algebra

Definition

Suppose that x, y and $\xi x + \eta y$ are elements of a Lie algebra $\mathcal{L}$ and that to every $x \in \mathcal{L}$ there exists a $d \times d$-matrix $\underline{\underline{R}}(x)$ such that

$$\underline{\underline{R}}(\xi x + \eta y) = \xi\underline{\underline{R}}(x) + \eta\underline{\underline{R}}(y) \text{ and} \tag{1.47}$$

$$\underline{\underline{R}}([x, y]) = \left[\underline{\underline{R}}(x), \underline{\underline{R}}(y)\right]. \tag{1.48}$$

Then these matrices are said to form a d-dimensional representation of $\mathcal{L}$.

Clearly, the set of matrices $\underline{\underline{R}}(x)$ forms a Lie algebra over the same field (with real or complex coefficients ξ, η) as $\mathcal{L}$.

In (1.21) the **adjoint** or ***ad* matrices** were introduced. Starting from (1.22) and (1.23) for simple Lie algebras it can be proved that the matrices $\{\underline{\underline{ad}}(x), \underline{\underline{ad}}(y), \ldots\}$ constitute a representation of the algebra $\{x, y, \ldots\}$. This is the *regular or adjoint representation.*

We come back to the elements $\Gamma(\boldsymbol{x})_{lk}$, (1.32) and (1.36), which make up the matrix $\underline{\underline{\Gamma}}(\boldsymbol{x})$. We now maintain that **the matrices $\underline{\underline{\Gamma}}(\boldsymbol{x})$, $\underline{\underline{\Gamma}}(\boldsymbol{y})$, ... constitute a representation of the Lie algebra $\boldsymbol{x}$,$\boldsymbol{y}$, ...** . As in Section 1.3 we presuppose that the set of basis functions $\psi_1, \psi_2, \ldots, \psi_d$ is associated with the Lie algebra and the relation (1.32) holds as $\boldsymbol{x}\,\psi_k = \sum\limits_{l=1}^{d} \Gamma(\boldsymbol{x})_{lk}\psi_l$ with $\Gamma(\boldsymbol{x})_{lk} = \langle \psi_l \mid \boldsymbol{x}\psi_k\rangle$, see (1.36). According to (1.35), the states $\psi_1,\ \psi_2, \ldots, \psi_d$ can be regarded as orthonormalized, i.e., $\langle \psi_l \mid \psi_m\rangle = \delta_{lm}$.

To prove our assertion is simple. First we deal with the linearity property of the matrices $\underline{\underline{\Gamma}}$. For the moment we handle with the matrix element (lk):

$$\Gamma(\xi\boldsymbol{x} + \eta\boldsymbol{y})_{lk} = \langle \psi_l \mid (\xi\boldsymbol{x} + \eta\boldsymbol{y})\,\psi_k\rangle = \xi\Gamma(\boldsymbol{x})_{lk} + \eta\Gamma(\boldsymbol{y})_{lk}\ . \tag{1.49}$$

Of course, the same relation holds for the entire matrices:

$$\underline{\underline{\Gamma}}(\xi\boldsymbol{x} + \eta\boldsymbol{y}) = \xi\underline{\underline{\Gamma}}(\boldsymbol{x}) + \eta\underline{\underline{\Gamma}}(\boldsymbol{y})\,. \tag{1.50}$$

Next we treat the commutator relation, which will be similar to (1.48). For the commutator $[\boldsymbol{x}, \boldsymbol{y}]$ we don't take the abstract form (1.2) but we choose

$$[\boldsymbol{x}, \boldsymbol{y}] = \boldsymbol{x}\boldsymbol{y} - \boldsymbol{y}\boldsymbol{x}. \tag{1.51}$$

Following (1.50) we write

$$\underline{\underline{\Gamma}}([\boldsymbol{x}, \boldsymbol{y}]) = \underline{\underline{\Gamma}}(\boldsymbol{x}\boldsymbol{y} - \boldsymbol{y}\boldsymbol{x}) = \underline{\underline{\Gamma}}(\boldsymbol{x}\boldsymbol{y}) - \underline{\underline{\Gamma}}(\boldsymbol{y}\boldsymbol{x}). \tag{1.52}$$

With eq. (1.42) we obtain

$$\underline{\underline{\Gamma}}([\boldsymbol{x}, \boldsymbol{y}]) = \underline{\underline{\Gamma}}(\boldsymbol{x})\underline{\underline{\Gamma}}(\boldsymbol{y}) - \underline{\underline{\Gamma}}(\boldsymbol{y})\underline{\underline{\Gamma}}(\boldsymbol{x}) = \left[\underline{\underline{\Gamma}}(\boldsymbol{x}), \underline{\underline{\Gamma}}(\boldsymbol{y})\right]. \tag{1.53}$$

Therefore, the equations (1.50) and (1.53) show that the matrices $\underline{\underline{\Gamma}}(\boldsymbol{x})$, $\underline{\underline{\Gamma}}(\boldsymbol{y})$, ... represent the Lie algebra $\{\boldsymbol{x}, \boldsymbol{y}, \ldots\}$.

1.5 Reducible and irreducible representations, multiplets

Let us suppose that the vector space of functions of a Lie algebra is the direct sum of two subspaces A and B, i.e., that the basis functions are split into two sets $\psi_1, \psi_2, \ldots, \psi_a \in A$ and $\psi_{a+1}, \psi_{a+2}, \ldots, (\psi_{a+b} = \psi_d) \in B$. Furthermore, we assume that the subspaces are *invariant*, i.e., for every operator$\boldsymbol{x}$ of the Lie algebra the relation (1.32) is modified like this:

$$\boldsymbol{x}\,\psi_k = \sum_{l=1}^{d} \Gamma(\boldsymbol{x})_{lk}\,\psi_l\,, \quad \begin{array}{l} \psi_l \in A \quad \text{if} \quad \psi_k \in A \quad \text{and} \\ \psi_l \in B \quad \text{if} \quad \psi_k \in B. \end{array} \tag{1.54}$$

That is, the transformation of the functions takes place only in the vector subspace of the original basis functions. This means that

$$\begin{array}{llll} \text{if} & 1 \le k \le a & \text{then} & 1 \le l \le a \quad \text{or} \\ \text{if} & a+1 \le k \le a+b=d & \text{then} & a+1 \le l \le a+b=d. \end{array} \tag{1.55}$$

Consequently, only those coefficients $\Gamma(\boldsymbol{x})_{lk}$ are non-zero which meet the conditions (1.55). Therefore the matrix $\underline{\underline{\Gamma}}(\boldsymbol{x})$ has the following form:

$$\underline{\underline{\Gamma}}(\boldsymbol{x}) = \begin{pmatrix} \Gamma_{11} & . & \Gamma_{1a} & 0 & 0 & . & 0 \\ . & . & . & . & . & . & . \\ \Gamma_{a1} & . & \Gamma_{aa} & . & . & . & 0 \\ 0 & . & . & \Gamma_{a+1,a+1} & . & . & \Gamma_{a+1,d} \\ 0 & . & . & . & . & . & . \\ . & . & . & . & . & . & . \\ 0 & . & 0 & \Gamma_{d.a+1} & . & . & \Gamma_{d,d} \end{pmatrix}. \tag{1.56}$$

If in a Lie algebra all matrices $\underline{\underline{\Gamma}}(\boldsymbol{x})$ are divided in this way in two or more squares along the diagonal, the representation is named *reducible.* Some authors call it "completely reducible".

On the other hand, if the operators constituting the algebra transform every function among functions of the entire vector space, there is *no nontrivial invariant subspace* (as defined at the beginning of this section). In this case, the matrices $\underline{\underline{\Gamma}}(\boldsymbol{x})$ cannot be decomposed in two or more square matrices along the diagonal, and the representation is named *irreducible.* The vector space of the functions of such a representation is called a *multiplet.* For particle physics, multiplets are very important.

We go back to the representation (1.56). The squares in the matrix are irreducible representations, i.e., the reducible representation is decomposed into the *direct sum of irreducible representations.*

If we interchange the basis functions or if we transform the basis more generally, it happens that we loose the structure of (1.56), i.e., the non-zero matrix elements can be spread over the entire matrix and the representation seems not to be reducible. However, one can reduce such a representation to the form (1.56) by a similarity transformation, where every matrix of the representation is transformed in the same way as set out below. *H. Weyl* proved that *every finite dimensional representation of a semi-simple algebra decomposes into the direct sum of irreducible representations.* Since the algebras $su(N)$ are simple, this theorem holds also for them.

We investigate the **transformation of representations**. If $\underline{\underline{S}}$ is a non-singular $d \times d$-matrix, starting from a representation of the Lie algebra with the d-dimensional matrices $\underline{\underline{\Gamma}}(\boldsymbol{x})$, we are able to construct a new representation constituted by the elements

$$\underline{\underline{\Gamma'}}(\boldsymbol{x}) = \underline{\underline{S}}^{-1}\underline{\underline{\Gamma}}(\boldsymbol{x})\,\underline{\underline{S}}. \tag{1.57}$$

The matrix $\underline{\underline{S}}$ is independent of $\mathbf{x}$.

First, we have to show that the matrices $\underline{\underline{\Gamma'}}(\boldsymbol{x})$ meet the linearity relation (1.50), namely

$$\underline{\underline{\Gamma'}}(\xi\boldsymbol{x}+\eta\boldsymbol{y}) = \underline{\underline{S}}^{-1}\underline{\underline{\Gamma}}(\xi\boldsymbol{x}+\eta\boldsymbol{y})\,\underline{\underline{S}} = \xi\underline{\underline{\Gamma'}}(\boldsymbol{x}) + \eta\underline{\underline{\Gamma'}}(\boldsymbol{y})\,. \tag{1.58}$$

Then, the commutator relation is treated like this:

$$\begin{aligned}\underline{\underline{\Gamma'}}([\boldsymbol{x},\boldsymbol{y}]) &= \underline{\underline{S}}^{-1}\underline{\underline{\Gamma}}(\boldsymbol{x})\,\underline{\underline{\Gamma}}(\boldsymbol{y})\,\underline{\underline{S}} - \underline{\underline{S}}^{-1}\underline{\underline{\Gamma}}(\boldsymbol{y})\,\underline{\underline{\Gamma}}(\boldsymbol{x})\,\underline{\underline{S}} \\ &= \underline{\underline{S}}^{-1}\underline{\underline{\Gamma}}(\boldsymbol{x})\,\underline{\underline{S}}^{-1}\underline{\underline{S}}\,\underline{\underline{\Gamma}}(\boldsymbol{y})\,\underline{\underline{S}} - \underline{\underline{S}}^{-1}\underline{\underline{\Gamma}}(\boldsymbol{y})\,\underline{\underline{S}}^{-1}\underline{\underline{S}}\,\underline{\underline{\Gamma}}(\boldsymbol{x})\,\underline{\underline{S}} \\ &= \left[\underline{\underline{\Gamma'}}(\boldsymbol{x})\,,\underline{\underline{\Gamma'}}(\boldsymbol{y})\right].\end{aligned} \tag{1.59}$$

Consequently, the set of matrices $\underline{\underline{\Gamma'}}(\boldsymbol{x})$ also forms a d-dimensional representation. The representations $\{\underline{\underline{\Gamma}}\}$ and $\{\underline{\underline{\Gamma'}}\}$ are said to be *equivalent.*

A given representation $\{\underline{\underline{\Gamma}}\}$ is reducible, if one can find the matrix $\underline{\underline{S}}$ in (1.57) so that all matrices $\underline{\underline{\Gamma}}'(\boldsymbol{x})$ of the equivalent representation have the same block-diagonal form (1.56) with two or more squares. If the matrices of the basis elements, $\underline{\underline{\Gamma}}'(\boldsymbol{e}_1), \underline{\underline{\Gamma}}'(\boldsymbol{e}_2), \ldots, \underline{\underline{\Gamma}}'(\boldsymbol{e}_n)$, have obtained this form, obviously every matrix of the equivalent representation shows the same form (Note: the commutator of two matrices with the same block diagonal structure generates a matrix which has the same structure).

Assume that the basis functions $\psi_1, \psi_2, \ldots, \psi_d$ are assigned to the representation $\{\underline{\underline{\Gamma}}(\boldsymbol{x})\}$. What can be said about the **basis functions** $\psi_1', \psi_2', \ldots, \psi_d'$ **which belong to the equivalent representation** $\{\underline{\underline{\Gamma}}'(\boldsymbol{x})\}$? We claim that the basis functions

$$\psi_n' = \sum_{m=1}^{d} S_{mn}\psi_m \quad \text{(with } \underline{\underline{S}} \text{ from (1.57))} \tag{1.60}$$

meet the relation $\boldsymbol{x}\psi_n' = \sum_{l=1}^{d} \Gamma'(\boldsymbol{x})_{ln}\psi_l'$. We let the linear operator$\boldsymbol{x}$ act on equation (1.60). Making use of (1.32) we get

$$\boldsymbol{x}\,\psi_n' = \sum_{m=1}^{d} S_{mn}\boldsymbol{x}\,\psi_m = \sum_{m,p=1}^{d} S_{mn}\Gamma(\boldsymbol{x})_{pm}\psi_p.$$

Because $\underline{\underline{S}}$ is non-singular, the equation (1.60) can be inverted like this $\psi_p = \sum_{q=1}^{d}\left(S^{-1}\right)_{qp}\psi_q'$, which we insert using (1.57) like this:

$$\begin{aligned}
\boldsymbol{x}\,\psi_n' &= \sum_{m,p,q=1}^{d} S_{mn}\Gamma(\boldsymbol{x})_{pm}\left(S^{-1}\right)_{qp}\psi_q' \\
&= \sum_{m,p,q=1}^{d} \left(S^{-1}\right)_{qp}\Gamma(\boldsymbol{x})_{pm}S_{mn}\,\psi_q' \qquad (1.61)\\
&= \sum_{q}^{d}\left(\underline{\underline{S}}^{-1}\underline{\underline{\Gamma}}(\boldsymbol{x})\,\underline{\underline{S}}\right)_{qn}\psi_q' = \sum_{q}^{d}\Gamma'(\boldsymbol{x})_{qn}\,\psi_q', \quad \text{QED.}
\end{aligned}$$

If the matrices $\underline{\underline{\Gamma}}'(\boldsymbol{x})$ of the equivalent representation have the form (1.56) with two or more squares on the diagonal, the vector space of the functions $\psi_1', \psi_2', \ldots, \psi_d'$ is divided in invariant subspaces. Due to (1.61) the operator $\boldsymbol{x}$ (and all the operators of the Lie algebra) transforms the states of a subspace among themselves as described in the example at the beginning of this section. The structure of the basis functions is given in (1.60).

Chapter 2

The Lie algebras $su(N)$

Square Hermitian matrices and their basis elements are displayed. The traceless version is formed. Square anti-Hermitian matrices are shown to constitute a Lie algebra ($u(N)$). The traceless version is $su(N)$. The Lie group $SU(N)$ is introduced and its correspondence to $su(N)$ is described. The total antisymmetry of the structure constants of $su(N)$ is shown. Now the Killing form can be given in more detail.

2.1 Hermitian matrices

Before we deal with $su\,(N)$-algebras, we look at Hermitian matrices $\underline{\underline{\lambda}}$ of the dimension N ($N \times N$-matrices). Hermiticity means that the adjoint of the matrix $\underline{\underline{\lambda}}$, i.e., the conjugate complex and transposed (reflected on the diagonal) matrix $\underline{\underline{\lambda}}^\dagger$ is identical with $\underline{\underline{\lambda}}$, that is,

$$\underline{\underline{\lambda}}^\dagger = \underline{\underline{\lambda}}. \tag{2.1}$$

This equation results in the following condition for matrix elements:

$$\lambda_{ik}^* = \lambda_{ki}. \tag{2.2}$$

If we have, for instance, $\lambda_{32} = r + \mathrm{i}s$ (r, s real, $\mathrm{i} = \sqrt{-1}$) and if all other matrix elements vanish apart from λ_{23}, the matrix reads

$$\begin{pmatrix} 0 & 0 & 0 & & 0 \\ 0 & 0 & r - \mathrm{i}s & & \\ 0 & r + \mathrm{i}s & 0 & & \\ & & & \cdot & \\ 0 & & & & 0 \end{pmatrix} = r \cdot \begin{pmatrix} 0 & 0 & 0 & & 0 \\ 0 & 0 & 1 & & \\ 0 & 1 & 0 & & \\ & & & \cdot & \\ 0 & & & & 0 \end{pmatrix} + s \cdot \begin{pmatrix} 0 & 0 & 0 & & 0 \\ 0 & 0 & -\mathrm{i} & & \\ 0 & \mathrm{i} & 0 & & \\ & & & \cdot & \\ 0 & & & & 0 \end{pmatrix}. \tag{2.3}$$

We observe that every matrix of the type (2.1) can be constructed linearly with the following basis elements

$$\begin{pmatrix} 1 & 0 & & & 0 \\ 0 & 0 & & & \\ & & \cdot & & \\ & & & \cdot & \\ 0 & & & & 0 \end{pmatrix}, \begin{pmatrix} 0 & 0 & & & 0 \\ 0 & 1 & & & \\ & & \cdot & & \\ & & & \cdot & \\ 0 & & & & 0 \end{pmatrix}, \ldots, \begin{pmatrix} 0 & 0 & & & 0 \\ 0 & 0 & & & \\ & & \cdot & & \\ & & & \cdot & \\ 0 & & & & 1 \end{pmatrix}, \tag{2.4}$$

$$\begin{pmatrix} 0 & 1 & & & 0 \\ 1 & 0 & & & \\ & & \cdot & & \\ & & & \cdot & \\ 0 & & & & 0 \end{pmatrix}, \begin{pmatrix} 0 & 0 & 1 & & 0 \\ 0 & 0 & & & \\ 1 & & 0 & & \\ & & & \cdot & \\ 0 & & & & 0 \end{pmatrix}, \ldots, \begin{pmatrix} 0 & 0 & & & 0 \\ 0 & 0 & & & \\ & & \cdot & & \\ & & & 0 & 1 \\ 0 & & & 1 & 0 \end{pmatrix},$$

$$\begin{pmatrix} 0 & -i & & & 0 \\ i & 0 & & & \\ & & \cdot & & \\ & & & \cdot & \\ 0 & & & & 0 \end{pmatrix}, \begin{pmatrix} 0 & 0 & -i & & 0 \\ 0 & 0 & & & \\ i & & 0 & & \\ & & & \cdot & \\ 0 & & & & 0 \end{pmatrix}, \ldots, \begin{pmatrix} 0 & 0 & & & 0 \\ 0 & 0 & & & \\ & & \cdot & & \\ & & & 0 & -i \\ 0 & & & i & 0 \end{pmatrix}, \tag{2.5}$$

if we admit only real coefficients in the construction. Obviously, the matrices (2.5) have vanishing traces (sums of the diagonal elements). If one replaces the matrices (2.4) by

$$\begin{pmatrix} 1 & 0 & & & 0 \\ 0 & -1 & & & \\ & & 0 & & \\ & & & \cdot & \\ 0 & & & & 0 \end{pmatrix}, \begin{pmatrix} 0 & 0 & & & 0 \\ 0 & 1 & & & \\ & & -1 & & \\ & & & \cdot & \\ 0 & & & & 0 \end{pmatrix}, \ldots, \begin{pmatrix} 0 & & & & 0 \\ & \cdot & & & \\ & & 0 & & \\ & & & 1 & \\ 0 & & & & -1 \end{pmatrix}, \tag{2.6}$$

one obtains a complete set of Hermitian basis matrices with vanishing traces. It is easy to show that an arbitrary diagonal and Hermitian matrix with vanishing trace can be constructed by a real linear combination of the matrices (2.6). In the literature, instead of the set (2.6), a set with the same number of linearly independent linear combinations of these matrices are used (see Section 2.3, (4.2), (5.3)). Simple counting reveals that there are N^2 matrices in (2.4) and (2.5). Replacing (2.4) by (2.6) reduces the number by 1 so that we obtain $N^2 - 1$. We will see that this number coincides with the dimension of the $su\,(N)$-algebra.

2.2 Definition

We maintain that **square anti-Hermitian matrices** of dimension N form a Lie algebra. A matrix $\underline{\underline{a}}$ is anti-Hermitian if its adjoint equals $-\underline{\underline{a}}$, that is,

$$\underline{\underline{a}}^\dagger = -\underline{\underline{a}}. \tag{2.7}$$

Taking real coefficients α and β, for two anti-Hermitian matrices $\underline{\underline{a}}$ and $\underline{\underline{b}}$ with the same dimension of course the following linear relation holds:

$$(\alpha\underline{\underline{a}} + \beta\underline{\underline{b}})^\dagger = \alpha\underline{\underline{a}}^\dagger + \beta\underline{\underline{b}}^\dagger = -\alpha\underline{\underline{a}} - \beta\underline{\underline{b}} = -\left(\alpha\underline{\underline{a}} + \beta\underline{\underline{b}}\right). \tag{2.8}$$

Thus, linear combinations with real coefficients of anti-Hermitian matrices are also anti-Hermitian. In order to investigate the commutator we make use of the well-known relation

$$\left(\underline{\underline{a}}\,\underline{\underline{b}}\right)^\dagger = \underline{\underline{b}}^\dagger\underline{\underline{a}}^\dagger \quad \text{and write} \tag{2.9}$$

$$\left[\underline{\underline{a}}, \underline{\underline{b}}\right]^\dagger = \left(\underline{\underline{ab}}\right)^\dagger - \left(\underline{\underline{ba}}\right)^\dagger = \underline{\underline{b}}^\dagger\underline{\underline{a}}^\dagger - \underline{\underline{a}}^\dagger\underline{\underline{b}}^\dagger = \underline{\underline{ba}} - \underline{\underline{ab}} = -\left[\underline{\underline{a}}, \underline{\underline{b}}\right]. \tag{2.10}$$

Obviously, the commutator of two anti-Hermitian matrices is also anti-Hermitian.

Confirming our assertion, these results show that anti-Hermitian $N \times N$-matrices constitute a *real Lie algebra* (with real coefficients). It is named $u\,(N)$-algebra because it is closely related to the Lie group of the unitary matrices. On the one hand, it constitutes a vector space over the field of real numbers and on the other hand, the matrices of the real matrix algebra $u\,(N)$ contain complex elements.

We go into the **elements of the matrix algebra *u(N)***. We substitute the general anti-Hermitian matrix by

$$\underline{\underline{a}} = -\frac{1}{2}\mathrm{i}\underline{\underline{h}} \tag{2.11}$$

and form the adjoint equation $\underline{\underline{a}}^\dagger = \frac{1}{2}\mathrm{i}\underline{\underline{h}}^\dagger$. Inserting both expressions in (2.7) we obtain

$$\frac{1}{2}\mathrm{i}\underline{\underline{h}}^\dagger = \frac{1}{2}\mathrm{i}\underline{\underline{h}}, \tag{2.12}$$

i.e., the matrix $\underline{\underline{h}}$ is Hermitian. As mentioned, the matrices (2.4) and (2.5),which we now name $\underline{\underline{\lambda}}_i$ $(i = 1, \ldots, N^2)$, are a basis for Hermitian matrices. Therefore, the elements $\left\{-\frac{1}{2}\mathrm{i}\lambda_i\right\}$ are a basis for the vector space of anti-Hermitian matrices, i.e., for $u\,(N)$. (By the way, note that the commutator of Hermitian matrices is not Hermitian but anti-Hermitian, which can be proven in analogy to (2.10)).

Here we look into the **direct sum of the real vector spaces *u(N)*** (set up by anti-Hermitian matrices) **and i*u(N)*** (Hermitian matrices). It comprises the basis elements $-\frac{1}{2}\mathrm{i}\underline{\underline{\lambda}}_i$ and $-\mathrm{i}\frac{1}{2}\mathrm{i}\underline{\underline{\lambda}}_i = \frac{1}{2}\underline{\underline{\lambda}}_i$ with $\underline{\underline{\lambda}}_i'$s from (2.4) and(2.5). Pairs of these elements form the $2N^2$ basis elements of (1.29). As explained there, their real vector space coincides with the vector space of $m\,(N, \mathbb{C})$. Thus we can write

$$m\,(N, \mathbb{C}) = u\,(N) \oplus \mathrm{i}u\,(N). \tag{2.13}$$

The symbol $\oplus$ denotes the direct sum. The vector spaces $u\,(N)$ and $\mathrm{i}u\,(N)$ are isomorphic and each has the dimension N^2. The vector space $u\,(N) \oplus \mathrm{i}u\,(N)$ can also be obtained starting with $u(N)$ by replacing its field of the real numbers by the field of the complex numbers. This procedure is named *complexification of the real algebra* $u(N)$. *A real Lie algebra can be complexified in case its basis elements stay linearly independent when the field of numbers is replaced by a complex one.* It can be shown that the Lie algebras $u(N)$ and $su(N)$ satisfy this condition.

Now we restrict ourselves to **anti-Hermitian $N \times N$-matrices a_s, b_s, ... with vanishing traces**. Of course, linear combinations of such matrices have also vanishing traces. Generally, the trace of the commutator of matrices vanishes because of the matrix relation $Tr\left(\underline{\underline{ab}}\right) = Tr\left(\underline{\underline{ba}}\right)$. Analogously to (2.11), we write $\underline{\underline{a_s}} = -\frac{1}{2}i\underline{\underline{h_s}}$ defining the corresponding Hermitian matrix $\underline{\underline{h_s}}$, which has also a vanishing trace:

$$Tr\left(\underline{\underline{h}}_s\right) = 0. \tag{2.14}$$

Every anti-Hermitian matrix with vanishing trace can be formed as a linear combination with real coefficients of the matrices

$$\underline{\underline{e}}_i = -\frac{1}{2}i\underline{\underline{\lambda}}_i \tag{2.15}$$

where $\underline{\underline{\lambda}}_i$ are basis matrices given in (2.6) and (2.5). Therefore, we have found that the basis elements $\underline{\underline{e}}_i$, (2.15), constitute a real Lie algebra. It represents a special kind of the $u(N)$ algebra and is named $su(N)$. It is a subalgebra of the real Lie algebra $u(N)$ and has $N^2 - 1$ basis matrices (see Section 2.1). According to point f) in Subsection 1.1.1, it possesses therefore the

$$\text{dimension} \quad n = N^2 - 1 \tag{2.16}$$

(not to confound with the matrix dimension N). Sometimes, the Hermitian matrices $\underline{\underline{\lambda}}_i$, which are given in (2.6) and (2.5), are named generators of the algebra $su(N)$. Note that they do not form a Lie algebra (the $\underline{\underline{e}}_i$'s do!).

It can be shown that *the Lie algebras $su(N)$ are simple* (see Subsection 1.1.5 and Section 1.5) In Section 4.2 we will check this property for $su(3)$.

Here we mention the **relation between the Lie algebra $su(N)$ and the compact Lie group $SU(N)$**. By definition, the group $SU(N)$ is constituted by unitary, N-dimensional matrices $\underline{\underline{A}}$ $\left(\underline{\underline{A}}^\dagger = \underline{\underline{A}}^{-1}\right)$ with determinant $\det \underline{\underline{A}} = 1$. The product $\underline{\underline{A}}\,\underline{\underline{B}}$ of arbitrary members $\underline{\underline{A}}$ and $\underline{\underline{B}}$ meets

$$\begin{aligned} \left(\underline{\underline{A}}\,\underline{\underline{B}}\right)^\dagger &= \underline{\underline{B}}^\dagger \underline{\underline{A}}^\dagger = \underline{\underline{B}}^{-1}\underline{\underline{A}}^{-1} = \left(\underline{\underline{A}}\,\underline{\underline{B}}\right)^{-1} \quad \text{and} \\ \det\left(\underline{\underline{A}}\,\underline{\underline{B}}\right) &= \det \underline{\underline{A}} \cdot \det \underline{\underline{B}} = 1, \end{aligned} \tag{2.17}$$

where well-known matrix relations have been applied. We see that the product is also unitary, has determinant 1 and is therefore a member of the group. For every $\underline{\underline{A}}$ exists the inverse element $\underline{\underline{A}}^{-1}$. The product $\underline{\underline{A}}\,\underline{\underline{A}}^{-1} = \underline{\underline{E}} \equiv \underline{\underline{1}}_N$ yields the identity element $\underline{\underline{E}}$, which meets $\underline{\underline{E}}\,\underline{\underline{A}} = \underline{\underline{A}}\,\underline{\underline{E}} = \underline{\underline{A}}$. These facts ensure that the elements $\underline{\underline{A}}, \underline{\underline{B}}, \ldots$ form a group. This group $SU(N)$ fulfils more sophisticated criteria not explained here which identify it to be a *linear and compact Lie group.*

It can be shown that the matrix elements of the $N \times N$-matrices $\underline{\underline{A}}$ of $SU(N)$ are analytic functions of n real parameters $x_1, x_2, \ldots, x_n$ with $n = N^2 - 1$, as we will display for $SU(2)$ in Section 3.1. The number n is the dimension of the group. The analytic dependence permits to differentiate the matrix elements $A_{ik}(x_1, x_2, \ldots, x_n)$ of the group element $\underline{\underline{A}}(x_1, x_2, \ldots, x_n)$ of $SU(N)$ with respect to x_l; i.e., the function $\frac{\partial}{\partial x_l} A_{ik}(x_1, x_2, \ldots, x_n)$ exists for every set of parameters $\{x_1, x_2, \ldots, x_n\}$.

Without proof, we state the following remarkable relation between the group $SU(N)$ and the algebra $su(N)$:

$$\frac{\partial}{\partial x_l} A_{ik}(x_1, x_2, \ldots, x_n)_{x_1 = x_2 = \ldots = x_n = 0} = (e_l)_{ik} \qquad (2.18)$$

$$(i, k = 1, 2 \ldots, N;\ l = 1, 2, \ldots, n)$$

where $(e_l)_{ik}$ is a matrix element of the basis element $\underline{\underline{e}}_l$ of the real Lie algebra $su(N)$. In summary one says: *the linear Lie group $SU(N)$ "corresponds" to the real Lie algebra $su(N)$*. This relation holds generally: *for every linear Lie group exists a corresponding real Lie algebra.* In Section 3.1 we will check this theorem for $SU(2)$ and $su(2)$.

The Lie algebra *su(N)* can also be **constituted by operators**. Their basis elements $\boldsymbol{e}_j$ correspond one-to-one to the basis matrices $\underline{\underline{e}}_j$ (see (2.15)). We define the operators $\boldsymbol{e}_j$ by the equation

$$\boldsymbol{e}_j \psi_k = \sum_{l=1}^{N} (e_j)_{lk} \psi_l. \qquad (2.19)$$

That is, we attribute a multiplet of N functions — the so-called fundamental multiplet — to the $su(N)$ -algebra and demand that the action of $\boldsymbol{e}_j$ on a function ψ_k generates the linear combination (2.19). The quantity $(e_j)_{lk}$ is a matrix element of $\underline{\underline{e}}_j$. The relation (2.19) corresponds to (1.32) and consequently,

$$\langle \psi_i \mid \boldsymbol{e}_j \psi_k \rangle = (e_j)_{ik} \qquad (2.20)$$

holds in analogy to (1.36). Due to (2.7), the basis matrices $\underline{\underline{e}}_j$ are anti-Hermitian: $\underline{\underline{e}}_j^\dagger = -\underline{\underline{e}}_j$, i.e., $\left(e_j^\dagger\right)_{ik} = -(e_j)_{ik}$. Following (2.20) and (1.45), we have $\left\langle \psi_i \,\middle|\, \boldsymbol{e}_j^\dagger \psi_k \right\rangle = -\langle \psi_i \mid \boldsymbol{e}_j \psi_k \rangle$, which yields

$$\boldsymbol{e}_j^\dagger = -\boldsymbol{e}_j, \qquad (2.21)$$

i.e., the basis operators are also anti-Hermitian. Corresponding to (2.15) we define the operators $\boldsymbol{\lambda}_i$ by

$$e_j = -\frac{1}{2}\mathrm{i}\boldsymbol{\lambda}_j. \tag{2.22}$$

The fundamental multiplet satisfies

$$\langle\psi_i \mid \boldsymbol{\lambda}_j\psi_k\rangle = (\lambda_j)_{ik}, \quad \text{and} \tag{2.23}$$

$$\boldsymbol{\lambda}_j^\dagger = \boldsymbol{\lambda}_j \quad \text{holds.} \tag{2.24}$$

An analogous consideration reveals that basis matrices $\underline{\underline{\Gamma}}(e_j)$ (see (1.32)) which correspond to any function set of $su(N)$ are anti-Hermitian.

2.3 Structure constants of *su(N)*

Due to (1.12) we can write

$$\left[\underline{\underline{e}}_i, \underline{\underline{e}}_k\right] = \sum_{l=1}^{n} C_{ikl}\underline{\underline{e}}_l \tag{2.25}$$

which results in

$$\left[\underline{\underline{\lambda}}_i, \underline{\underline{\lambda}}_k\right] = \sum_{l=1}^{n} C_{ikl}2\mathrm{i}\underline{\underline{\lambda}}_l \tag{2.26}$$

by means of (2.15). The scalar C_{ikl} is a structure constant of the real Lie algebra $su(N)$ with basis elements $\underline{\underline{e}}_l$ (not $\underline{\underline{\lambda}}_l$!), and n is the dimension of the algebra. Because the algebra $su(N)$ is real, the structure constants are real (otherwise the right hand side of (2.25) would not be purely anti-Hermitian). In order to display further properties of C_{ikl}, we look at the trace of products of matrices $\underline{\underline{\lambda}}_i$. For Hermitian basis matrices in (2.5), say $\underline{\underline{\lambda}}_i$ and $\underline{\underline{\lambda}}_k$, the following relation holds:

$$Tr\left(\underline{\underline{\lambda}}_i\underline{\underline{\lambda}}_k\right) = 2\,\delta_{ik}, \tag{2.27}$$

which can be seen directly by treating products numerically. For the diagonal matrices in (2.6) this relation is only true for $su(2)$ but for the higher $su(N)$ it is not. However, the set of matrices (2.6) can be replaced by a set with the same number of linearly independent matrices which are real linear combinations of the matrices (2.6) and which satisfy the relation (2.27). For $su(3)$ this set is given in Section 4.1 and for $su(4)$ in Section 5.1. From now on, for the Hermitian, diagonal basis matrices $\underline{\underline{\lambda}}_i$ we take these ones and the off-diagonal matrices from (2.5).

Making use of (2.26) and (2.27), we calculate the trace

$$Tr\left(\left[\underline{\underline{\lambda}}_i, \underline{\underline{\lambda}}_k\right]\underline{\underline{\lambda}}_l\right) = 2\mathrm{i}\sum_m C_{ikm}Tr\left(\underline{\underline{\lambda}}_m\underline{\underline{\lambda}}_l\right) = 2\mathrm{i}\sum_m C_{ikm}2\delta_{ml} = 4\mathrm{i}C_{ikl}. \tag{2.28}$$

That is, the entire set of $\underline{\underline{\lambda}}_i$ given, the structure constants can be calculated by

$$C_{ikl} = \frac{1}{4\mathrm{i}} Tr\left(\left[\underline{\underline{\lambda}}_i, \underline{\underline{\lambda}}_k\right] \underline{\underline{\lambda}}_l\right). \tag{2.29}$$

Using again the matrix relation $Tr\left(\underline{\underline{ab}}\right) = Tr\left(\underline{\underline{ba}}\right)$, we can treat (2.28) as follows:

$$\begin{aligned} 4\mathrm{i}C_{ikl} &= Tr\left(\left[\underline{\underline{\lambda}}_i, \underline{\underline{\lambda}}_k\right] \underline{\underline{\lambda}}_l\right) = Tr\left(\underline{\underline{\lambda}}_i \underline{\underline{\lambda}}_k \underline{\underline{\lambda}}_l - \underline{\underline{\lambda}}_k \underline{\underline{\lambda}}_i \underline{\underline{\lambda}}_l\right) \\ &= -Tr\left(\underline{\underline{\lambda}}_k \underline{\underline{\lambda}}_i \underline{\underline{\lambda}}_l - \underline{\underline{\lambda}}_i \underline{\underline{\lambda}}_k \underline{\underline{\lambda}}_l\right) = -Tr\left(\underline{\underline{\lambda}}_i \underline{\underline{\lambda}}_l \underline{\underline{\lambda}}_k - \underline{\underline{\lambda}}_l \underline{\underline{\lambda}}_i \underline{\underline{\lambda}}_k\right) \\ &= -Tr\left(\left[\underline{\underline{\lambda}}_i, \underline{\underline{\lambda}}_l\right] \underline{\underline{\lambda}}_k\right) = -4\mathrm{i}C_{ilk}. \end{aligned} \tag{2.30}$$

In view of (1.20), we realize that every odd permutation of the indices of C_{ikl} changes its sign. In other words, C_{ikl} is totally antisymmetric in all indices. Consequently, no index appears more than once in a non-vanishing structure constant.

Here the **"Killing form"** corresponding to the basis elements $\underline{\underline{e}}_i$ and $\underline{\underline{e}}_j$ can be remodelled. Using (1.28) and (2.30), we write

$$B\left(\underline{\underline{e}}_i, \underline{\underline{e}}_j\right) = \sum_{l,k=1}^{n} C_{ikl} C_{jlk} = \sum_{l,k=1}^{n} -C_{ikl} C_{jkl}. \tag{2.31}$$

Therefore, the diagonal elements

$$B\left(\underline{\underline{e}}_i, \underline{\underline{e}}_i\right) = -\sum_{l,k=1}^{n} C_{ikl}^2 \tag{2.32}$$

are negative, because the structure constants are real as stated above. The trace of the matrix $\underline{\underline{B}}$, which contains the matrix elements $B\left(\underline{\underline{e}}_i, \underline{\underline{e}}_i\right)$, is negative. In Section 3.1 the "Killing form" of $su\,(2)$ is calculated in detail.

Chapter 3

The Lie algebra $su(2)$

The basis elements of the matrix algebra $su(2)$ and the corresponding structure constants are given. In detail it is shown that the Lie group $SU(2)$ "corresponds" to the algebra $su(2)$. Another detailed calculation yields the basis matrices of the adjoint representation of $su(2)$. They are used to make up the Killing form of $su(2)$.

Starting from the basis operators constituting $su(2)$, a convenient representation is looked for with a diagonal "third" element. Its diagonal values characterise its eigenfunctions. These are also eigenfunctions of the quadratic Casimir operator. The sets of eigenvalues of both operators are developed. Acting on these eigenfunctions, the basis operators — or their Hermitian equivalents — generate expansion coefficients, which result in irreducible representations. Their basis matrices are calculated for $j = 1/2$, 1 and 3/2.

The "direct product" of two irreducible representations and its affiliated product functions are investigated. An equivalent representation has affiliated functions, which are constructed using Clebsch–Gordan coefficients and which again set up eigenfunctions of the $su(2)$-operators. The representation matrices for $(j, j') = (1/2,1/2)$ and $(1,1/2)$ are calculated.

A graphical procedure is displayed which generates the set of multiplets resulting from a direct product of $su(2)$-multiplets.

3.1 The generators of the $su(2)$-algebra

From Section 2.2 we infer that anti-Hermitian 2×2-matrices with vanishing trace constitute the real Lie algebra $su(2)$. Following (2.15), (2.5) and (2.6), we write the basis elements like this:

$$\underline{\underline{e}}_1 = -\frac{1}{2}\mathrm{i}\begin{pmatrix} 0 & 1 \\ 1 & 0 \end{pmatrix}, \quad \underline{\underline{e}}_2 = -\frac{1}{2}\mathrm{i}\begin{pmatrix} 0 & -\mathrm{i} \\ \mathrm{i} & 0 \end{pmatrix}, \quad \underline{\underline{e}}_3 = -\frac{1}{2}\mathrm{i}\begin{pmatrix} 1 & 0 \\ 0 & -1 \end{pmatrix}. \tag{3.1}$$

These expressions show that $su(2)$ has dimension $3 = 2^2 - 1$ according to (2.16). The commutators read

$$\left[\underline{\underline{e}}_1, \underline{\underline{e}}_2\right] = \underline{\underline{e}}_3, \quad \text{cyclic in } 1,2,3. \tag{3.2}$$

Other exchanges of indices change the relative sign in (3.2), and no index appears twice in the same non-vanishing equation. From (2.25) and (3.2) we obtain the following simple structure constants:

$$C_{123} = C_{231} = C_{312} = 1, \tag{3.3}$$

which is also true for the isomorphic $su(2)$-algebra constituted by the basis operators $\boldsymbol{e}_i$, (2.16). Due to $\underline{\underline{e}}_i = -\frac{1}{2}\mathrm{i}\underline{\underline{\lambda}}_i$, (2.15), the basis elements can be written with the Hermitian matrices (generators)

$$\underline{\underline{\lambda}}_1 = \begin{pmatrix} 0 & 1 \\ 1 & 0 \end{pmatrix}, \quad \underline{\underline{\lambda}}_2 = \begin{pmatrix} 0 & -\mathrm{i} \\ \mathrm{i} & 0 \end{pmatrix}, \quad \underline{\underline{\lambda}}_3 = \begin{pmatrix} 1 & 0 \\ 0 & -1 \end{pmatrix}, \tag{3.4}$$

for which the relation

$$\left[\underline{\underline{\lambda}}_1, \underline{\underline{\lambda}}_2\right] = 2\mathrm{i}\underline{\underline{\lambda}}_3 \tag{3.5}$$

holds with cyclic permutations (see also (2.26)). The $\underline{\underline{\lambda}}_i$'s are named *Pauli spin matrices* and form a complex Lie algebra of 2×2-matrices.

At this early stage we mention the **parallelism between the real Lie algebra *su*(2) and the special, linear, complex algebra *sl*(2,ℂ)**. It comprises 2×2-matrices, and the term "special" denotes the vanishing traces. A natural basis of this algebra reads

$$\underline{\underline{x}} = \begin{pmatrix} 0 & 1 \\ 0 & 0 \end{pmatrix}, \quad \underline{\underline{y}} = \begin{pmatrix} 0 & 0 \\ 1 & 0 \end{pmatrix}, \quad 2\underline{\underline{z}} = \begin{pmatrix} 1 & 0 \\ 0 & -1 \end{pmatrix} \tag{3.6}$$

satisfying

$$\left[\underline{\underline{z}}, \underline{\underline{x}}\right] = \underline{\underline{x}}, \quad \left[\underline{\underline{z}}, \underline{\underline{y}}\right] = -\underline{\underline{y}}, \quad \left[\underline{\underline{x}}, \underline{\underline{y}}\right] = 2\underline{\underline{z}}. \tag{3.7}$$

Obviously, every 2×2-matrix (including complex elements) with vanishing trace can be written as a linear combination of the matrices (3.6) using complex coefficients. The coincidence of (3.7) with (3.43) identifies the algebra $\{J_+, J_-, J_3\}$ to be $sl(2,\mathbb{C})$. The remark in context with (3.69) reveals that $su(2)$ and $sl(2,\mathbb{C})$ have the same set of multiplets.

Here we prove **the correspondence between the Lie group *SU*(2) and the Lie algebra *su*(2)** as formulated in Section 2.2. An element of the group $SU(2)$, i.e., a unitary 2 × 2-matrix $\underline{\underline{U}}$ with det $\underline{\underline{U}} = 1$, can be written using the complex numbers α and β such that $|\alpha|^2 + |\beta|^2 = 1$ as follows:

$$\underline{\underline{U}} = \begin{pmatrix} \alpha & \beta \\ -\beta^* & \alpha^* \end{pmatrix}. \tag{3.8}$$

Obviously $\underline{\underline{U}}\,\underline{\underline{U}}^\dagger = \underline{\underline{1}}_2 \equiv \begin{pmatrix} 1 & 0 \\ 0 & 1 \end{pmatrix}$ and $\det \underline{\underline{U}} = 1$ as demanded. With $\alpha = \alpha_1 + \mathrm{i}\alpha_2$ and $\beta = \beta_1 + \mathrm{i}\beta_2$, $(\alpha_1, \alpha_2, \beta_1, \beta_2$ real), the determinant condition reads $\alpha_1^2 + \alpha_2^2 + \beta_1^2 + \beta_2^2 = 1$, i.e., there are three free parameters. We set conveniently $\alpha_2 = -\frac{1}{2}x_3$, $\beta_1 = -\frac{1}{2}x_2$, $\beta_2 = -\frac{1}{2}x_1$ and obtain $\alpha_1^2 = 1 - \frac{1}{4}\left(x_1^2 + x_2^2 + x_3^2\right)$. We take the minus sign for α_1. That is,

$$\underline{\underline{U}} = \begin{pmatrix} -\sqrt{1 - \frac{1}{4}\left(x_1^2 + x_2^2 + x_3^2\right)} - \frac{1}{2}\mathrm{i}x_3 & -\frac{1}{2}x_2 - \frac{1}{2}\mathrm{i}x_1 \\ \frac{1}{2}x_2 - \frac{1}{2}\mathrm{i}x_1 & -\sqrt{1 - \frac{1}{4}\left(x_1^2 + x_2^2 + x_3^2\right)} + \frac{1}{2}\mathrm{i}x_3 \end{pmatrix}. \tag{3.9}$$

We see that every matrix element of $\underline{\underline{U}}$ is an analytic function of the parameters x_1, x_2, x_3, as indicated in Section 2.2. Following (2.18) we calculate

$$\begin{aligned} \frac{\partial}{\partial x_1}\left(U_{11}\right)_{x_1=x_2=x_3=0} &= \frac{\partial}{\partial x_1}\left(U_{22}\right)_{x_1=x_2=x_3=0} = 0, \\ \frac{\partial}{\partial x_1}\left(U_{12}\right)_{x_1=x_2=x_3=0} &= \frac{\partial}{\partial x_1}\left(U_{21}\right)_{x_1=x_2=x_3=0} = -\frac{1}{2}\mathrm{i}. \end{aligned}$$

Using (3.1) we sum up: $\frac{\partial}{\partial x_1}\left(U_{ik}\right)_{x_1=x_2=x_3=0} = \left(e_1\right)_{ik}$ $(i, k = 1, 2)$. Analogously, one finds $\frac{\partial}{\partial x_2}\left(U_{ik}\right)_{x_1=x_2=x_3=0} = \left(e_2\right)_{ik}$ and $\frac{\partial}{\partial x_3}\left(U_{ik}\right)_{x_1=x_2=x_3=0} = \left(e_3\right)_{ik}$ in accordance with (2.18). Indeed, the group $SU(2)$ "corresponds" with the algebra $su(2)$.

How can a given Lie algebra with basis elements e_1', e_2' and e_3' be **identified as *su*(2)-algebra**? We clear up if every e_1', e_2' , e_3' can be written using real coefficients as a linear combination of the basis elements e_1, e_2 , e_3 of $su(2)$. The structure constants of the unidentified Lie algebra meet the relations

$$\sum_l C_{ikl}' e_l' = \left[e_i', e_k'\right]. \tag{3.10}$$

The ansatz just mentioned reads

$$e_p' = \sum_q d_{pq} e_q \quad , d_{pq} \text{ real}, \tag{3.11}$$

which we insert in (3.10) this way:

$$\sum_l C_{ikl}' d_{ln} e_n = \sum_{j,m} d_{ij} d_{km} \left[e_j, e_m\right] = \sum_{j\,m\,n} d_{ij} d_{km} C_{jmn} e_n. \tag{3.12}$$

Because the basis elements e_n are linearly independent, we obtain three equations (for n = 1,2,3): $\sum\limits_l C'_{ikl} d_{ln} = \sum\limits_{j\,m} d_{ij} d_{km} C_{jmn}$. With the values for C_{jmn} from (3.3), we have

$$\begin{aligned}\sum_l C'_{ikl} d_{l1} &= d_{i2} d_{k3} - d_{i3} d_{k2} \\ \sum_l C'_{ikl} d_{l2} &= d_{i3} d_{k1} - d_{i1} d_{k3} \quad . \\ \sum_l C'_{ikl} d_{l3} &= d_{i1} d_{k2} - d_{i2} d_{k1} \; .\end{aligned} \tag{3.13}$$

Effectively (3.13) displays 9 homogenic, second-order equations (for $i, k = 1, 2, 3$, $i < k$) for 9 quantities. It is possible that the system is not soluble. If a solution exists the investigated algebra is isomorphic to $su(2)$.

Here we look at the **regular or adjoint representation of $su(2)$** . Owing to (1.25), its basis elements consist of 3×3-matrices $\underline{\underline{ad}}\,(e_i)$, which contain the matrix elements

$$ad\,(e_i)_{kl} = C_{ikl}. \tag{3.14}$$

With (3.3) we obtain

$$\begin{aligned}\underline{\underline{ad}}\,(e_1) &= \begin{pmatrix} C_{111} & C_{121} & C_{131} \\ C_{112} & C_{122} & C_{132} \\ C_{113} & C_{123} & C_{133} \end{pmatrix} = \begin{pmatrix} 0 & 0 & 0 \\ 0 & 0 & -1 \\ 0 & 1 & 0 \end{pmatrix}, \\ \underline{\underline{ad}}\,(e_2) &= \begin{pmatrix} 0 & 0 & 1 \\ 0 & 0 & 0 \\ -1 & 0 & 0 \end{pmatrix}, \quad \underline{\underline{ad}}\,(e_3) = \begin{pmatrix} 0 & -1 & 0 \\ 1 & 0 & 0 \\ 0 & 0 & 0 \end{pmatrix}.\end{aligned} \tag{3.15}$$

Numerical calculations confirm the relation

$$\left[\underline{\underline{ad}}\,(e_1)\,,\underline{\underline{ad}}\,(e_2)\right] = \underline{\underline{ad}}\,(e_3) \tag{3.16}$$

with cyclic permutations, in analogy to (3.2). It results from the fact that the $\underline{\underline{ad}}$-matrices represent the algebra (see Section 1.4).

Here we calculate the **"Killing form" of $su(2)$** corresponding to the basis elements $\underline{\underline{e}}_i$ and $\underline{\underline{e}}_k$. This form is also named *metric tensor*. Owing to (2.31) we have

$$B\left(\underline{\underline{e}}_i, \underline{\underline{e}}_k\right) = \sum_{l,\,m=1}^{3} C_{ilm} C_{kml}, \qquad i, k = 1, 2, 3\,. \tag{3.17}$$

For $su(2)$ we obtain, for example, $B\left(\underline{\underline{e}}_1, \underline{\underline{e}}_2\right) = C_{123}C_{232} + C_{132}C_{223} + 0 = 0$ because of $C_{232} = C_{223} = 0$ (see (2.30) and (3.3)). So all off-diagonal elements of the matrix $\underline{\underline{B}}$ vanish:

$$B\left(\underline{\underline{e}}_i, \underline{\underline{e}}_k\right) = 0 \quad \text{for } i \neq k. \tag{3.18}$$

Analogously we have, for instance, $B\left(\underline{\underline{e}}_3, \underline{\underline{e}}_3\right) = C_{312}C_{321} + C_{321}C_{312} = -2$. In fact, every diagonal matrix element reads

$$B\left(\underline{\underline{e}}_i, \underline{\underline{e}}_i\right) = -2. \tag{3.19}$$

That is,

$$\underline{\underline{B}} = -2 \cdot \underline{\underline{1}}_3 \tag{3.20}$$

for $su(2)$ with
$$\underline{\underline{1}}_3 \equiv \begin{pmatrix} 1 & 0 & 0 \\ 0 & 1 & 0 \\ 0 & 0 & 1 \end{pmatrix}.$$

3.2 Operators constituting the algebra $su(2)$

The basis operators $\boldsymbol{e}_i$ were defined in (2.19) by

$$\boldsymbol{e}_i\,\psi_k = \sum_{l=1}^{2} (e_i)_{lk}\psi_l. \tag{3.21}$$

In fact, due to (3.1) and (3.21) we have

$$\boldsymbol{e}_1\psi_1 = -\frac{1}{2}\mathrm{i}\,(0+\psi_2)\,, \quad \boldsymbol{e}_1\psi_2 = -\frac{1}{2}\mathrm{i}\,(\psi_1+0)\,, \tag{3.22}$$

$$\boldsymbol{e}_2\psi_1 = -\frac{1}{2}\mathrm{i}\,(0+\mathrm{i}\psi_2)\,, \quad \boldsymbol{e}_2\psi_2 = -\frac{1}{2}\mathrm{i}\,(-\mathrm{i}\psi_1+0)\,, \tag{3.23}$$

$$\boldsymbol{e}_3\psi_1 = -\frac{1}{2}\mathrm{i}\,(\psi_1+0)\,, \quad \boldsymbol{e}_3\psi_2 = -\frac{1}{2}\mathrm{i}\,(0-\psi_2)\,. \tag{3.24}$$

Of course, the basis operators of $su(2)$ meet

$$[\boldsymbol{e}_1, \boldsymbol{e}_2] = \boldsymbol{e}_3\,, \quad \text{cyclic in } 1,2,3, \tag{3.25}$$

in analogy with (3.2). Because, making use of (2.20) and (1.41), the matrix relation (3.2) can be deduced from (3.25). The inverse is also true. Alluding to the quantum mechanical operator of the angular momentum, we introduce

$$\boldsymbol{e}_i = -\mathrm{i}\boldsymbol{J}_i. \tag{3.26}$$

From (2.21) we have

$$-\boldsymbol{e}_i = +\mathrm{i}\boldsymbol{J}_i = (-\mathrm{i}\boldsymbol{J}_i)^\dagger = +\mathrm{i}\boldsymbol{J}_i^\dagger. \tag{3.27}$$

That is, the operators $\boldsymbol{J}_i$ are Hermitian corresponding to $\boldsymbol{\lambda}_i$, (2.24). From (3.25) and (3.26) we obtain

$$[\boldsymbol{J}_1, \boldsymbol{J}_2] = \mathrm{i}\boldsymbol{J}_3 \quad \text{with cyclic permutations.} \tag{3.28}$$

We introduce the so-called quadratic Casimir operator of $su(2)$ (see also Section 4.11)

$$\boldsymbol{J}^2 = \boldsymbol{J}_1^2 + \boldsymbol{J}_2^2 + \boldsymbol{J}_3^2 \tag{3.29}$$

and look at the commutator

$$\begin{aligned}\left[\boldsymbol{J}^2, \boldsymbol{J}_1\right] = \left[\boldsymbol{J}_2^2, \boldsymbol{J}_1\right] + \left[\boldsymbol{J}_3^2, \boldsymbol{J}_1\right] &= \boldsymbol{J}_2\boldsymbol{J}_2\boldsymbol{J}_1 - \boldsymbol{J}_1\boldsymbol{J}_2\boldsymbol{J}_2 + \boldsymbol{J}_3\boldsymbol{J}_3\boldsymbol{J}_1 - \boldsymbol{J}_1\boldsymbol{J}_3\boldsymbol{J}_3 \\ &= \boldsymbol{J}_2\boldsymbol{J}_1\boldsymbol{J}_2 - \boldsymbol{J}_2\mathrm{i}\boldsymbol{J}_3 - \boldsymbol{J}_2\boldsymbol{J}_1\boldsymbol{J}_2 - \mathrm{i}\boldsymbol{J}_3\boldsymbol{J}_2 + \boldsymbol{J}_3\boldsymbol{J}_1\boldsymbol{J}_3 - \boldsymbol{J}_3\mathrm{i}\boldsymbol{J}_2 \\ &\quad - \boldsymbol{J}_3\boldsymbol{J}_1\boldsymbol{J}_3 + \mathrm{i}\boldsymbol{J}_2\boldsymbol{J}_3 = 0,\end{aligned} \tag{3.30}$$

which holds for every $\boldsymbol{J}_i$:

$$\left[\boldsymbol{J}^2, \boldsymbol{J}_i\right] = 0. \tag{3.31}$$

Due to (1.46), (3.27) and (3.29), $\boldsymbol{J}^2$ is Hermitian, too.

Before we investigate the eigenvalues of $\boldsymbol{J}^2$ and $\boldsymbol{J}_3$, we deal with the **representation of the algebra** $\{\mathbf{e}_1, \mathbf{e}_2, \mathbf{e}_3\}$. If an operator $\mathbf{e}_i$ acts on an arbitrary, affiliated function ψ_k, according to (1.32) the relation

$$\mathbf{e}_i\,\psi_k = \sum_{l=1}^{d} \Gamma\left(\mathbf{e}_i\right)_{lk}\psi_l \tag{3.32}$$

must hold. If the matrix $\underline{\underline{\Gamma}}(\mathbf{e}_3)$ is not diagonal, in order to achieve analogy to (3.1) and (2.19) it is diagonalized (it can be shown that the criteria for the diagonalizability are fulfilled). The well-known procedure has been displayed partly in Section 1.5, thus, here we outline it only briefly.

A non-singular matrix $\underline{\underline{S}}$ has to be found which forms an equivalent representation with the elements

$$\underline{\underline{\Gamma}}'(\mathbf{e}_i) = \underline{\underline{S}}^{-1}\underline{\underline{\Gamma}}(\mathbf{e}_i)\,\underline{\underline{S}}, \quad i = 1,\,2,\,3, \tag{3.33}$$

(see (1.57)) in such a way that $\underline{\underline{\Gamma}}'(\mathbf{e}_3)$ is diagonal with matrix elements $\gamma_1, \gamma_2, \ldots, \gamma_d$. From $\underline{\underline{S}}\,\underline{\underline{\Gamma}}'(\mathbf{e}_3) = \underline{\underline{\Gamma}}(\mathbf{e}_3)\,\underline{\underline{S}}$ we obtain for the matrix element (k, m):

$$\sum_{l=1}^{d} S_{kl}\Gamma'\left(\mathbf{e}_3\right)_{lm} = S_{km}\gamma_m = \sum_{l=1}^{d} \Gamma\left(\mathbf{e}_3\right)_{kl} S_{lm}. \tag{3.34}$$

For a given m we have a system of d linear, homogeneous equations

$$\begin{array}{rcl}(\Gamma(\mathbf{e}_3)_{11} - \gamma_m)\,S_{1m} + \quad \Gamma(\mathbf{e}_3)_{12}\,S_{2m} + \quad \Gamma(\mathbf{e}_3)_{13}\,S_{3m} + \ldots & = & 0 \\ \Gamma(\mathbf{e}_3)_{21}\,S_{1m} + (\Gamma(\mathbf{e}_3)_{22} - \gamma_m)S_{2m} + \quad \Gamma(\mathbf{e}_3)_{23}\,S_{3m} + \ldots & = & 0 \\ \Gamma(\mathbf{e}_3)_{31}\,S_{1m} + \quad \Gamma(\mathbf{e}_3)_{32}\,S_{2m} + (\Gamma(\mathbf{e}_3)_{33} - \gamma_m)S_{3m} + \ldots & = & 0 \\ \cdot\quad\cdot & \cdot & \cdot \\ \cdot\quad\cdot & \cdot & \cdot\;, \end{array} \tag{3.35}$$

which is soluble if the determinant of the system vanishes, i.e.,

$$\left|\underline{\underline{\Gamma}}(e_3) - \underline{\underline{1}}\gamma_m\right| = 0. \tag{3.36}$$

This secular equation yields d roots γ_m, which constitute $\underline{\underline{\Gamma'}}(e_3)$ and which permit to calculate the matrix $\underline{\underline{S}}$ by means of (3.35). According to (1.60), the new basis functions $|k\rangle$, which correspond to the representation $\left\{\underline{\underline{\Gamma'}}(e_i)\right\}$, read

$$|k\rangle = \sum_m S_{mk}\,\psi_m. \tag{3.37}$$

Because $\underline{\underline{\Gamma'}}(e_3)$ is diagonal, according to (1.61), the following eigenvalue equations hold:

$$e_3\,|k\rangle = \gamma_k\,|k\rangle\,, \quad k = 1,\,2,\,\ldots,\,d. \tag{3.38}$$

Here we go back to the **operators $\boldsymbol{J}_3$ and $\boldsymbol{J}^2$**. We define $\gamma_k = -\mathrm{i}J_3^{(k)}$ and write with the aid of (3.26) $-\mathrm{i}\boldsymbol{J}_3\,|k\rangle = -\mathrm{i}J_3^{(k)}\,|k\rangle$. Since the state $|k\rangle$ is characterised by the eigenvalue $J_3^{(k)}$, we write for it $\left|J_3^{(k)}\right\rangle$. Finally, we drop the index k and obtain the following eigenvalue equation:

$$\boldsymbol{J}_3\,|J_3\rangle = J_3\,|J_3\rangle\,. \tag{3.39}$$

It can be shown that the eigenvalues J_3 are distinct. We now claim that the states $|J_3\rangle$ are also eigenstates of the operator $\boldsymbol{J}^2$, (3.29). Due to (3.31) we have $0 = \left[\boldsymbol{J}^2, \boldsymbol{J}_3\right]|J_3\rangle = \boldsymbol{J}^2\boldsymbol{J}_3\,|J_3\rangle - \boldsymbol{J}_3\boldsymbol{J}^2\,|J_3\rangle = (J_3\boldsymbol{J}^2 - \boldsymbol{J}_3\boldsymbol{J}^2)\,|J_3\rangle$ and $\boldsymbol{J}_3(\boldsymbol{J}^2\,|J_3\rangle) = J_3(\boldsymbol{J}^2\,|J_3\rangle)$. That is, $(\boldsymbol{J}^2\,|J_3\rangle)$ is eigenket of $\boldsymbol{J}_3$ alike to $|J_3\rangle$ and therefore it must be proportional to $|J_3\rangle$ like this:

$$\boldsymbol{J}^2\,|J_3\rangle = \bar{J}\,|J_3\rangle\,. \tag{3.40}$$

This means that $|J_3\rangle$ is also eigenfunction of $\boldsymbol{J}^2$ and is characterised by both eigenvalues, that is why we replace $|J_3\rangle$ by $\left|\bar{J}J_3\right\rangle$. The eigenvalues J_3 and $\bar{J}$ of the operators $\boldsymbol{J}_3$ and $\boldsymbol{J}^2$ are real, as is well-known for Hermitian operators.

3.3 Multiplets of $su(2)$

In order to investigate the quantities J_3 and $\bar{J}$ we introduce the following step operators (sometimes named ladder- or shift operators)

$$\boldsymbol{J}_+ = \boldsymbol{J}_1 + \mathrm{i}\boldsymbol{J}_2\,, \quad \boldsymbol{J}_- = \boldsymbol{J}_1 - \mathrm{i}\boldsymbol{J}_2. \tag{3.41}$$

They differ slightly from analogous operators used for example by Talmi, 1993, p. 79 or Pfeifer, 1998, p. 43. Here we follow partially the treatise of Biedenharn et. al., 1981, p. 31. Due to (3.27) the adjoint operator of $\boldsymbol{J}_+$ reads

$$\boldsymbol{J}_+^\dagger = (\boldsymbol{J}_1 + \mathrm{i}\boldsymbol{J}_2)^\dagger = \boldsymbol{J}_1^\dagger - \mathrm{i}\boldsymbol{J}_2^\dagger = \boldsymbol{J}_1 - \mathrm{i}\boldsymbol{J}_2 = \boldsymbol{J}_-\,. \tag{3.42}$$

Using the expressions (3.28), one proves the following commutator relations

$$[\boldsymbol{J}_+, \boldsymbol{J}_-] = 2\boldsymbol{J}_3\,, \quad [\boldsymbol{J}_3, \boldsymbol{J}_\pm] = \pm\boldsymbol{J}_\pm\,, \quad [\boldsymbol{J}^2, \boldsymbol{J}_\pm] = 0\,, \quad [\boldsymbol{J}^2, \boldsymbol{J}_3] = 0\,. \tag{3.43}$$

We define the following kets:

$$|+\rangle \equiv \boldsymbol{J}_+ \left|\bar{J}J_3\right\rangle \text{ and } |-\rangle \equiv \boldsymbol{J}_- \left|\bar{J}J_3\right\rangle, \tag{3.44}$$

and apply the third expression of (3.43): $0 = \left[\boldsymbol{J}^2, \boldsymbol{J}_\pm\right] \left|\bar{J}J_3\right\rangle = \boldsymbol{J}^2 \left|\pm\right\rangle - \boldsymbol{J}_\pm \bar{J} \left|\bar{J}J_3\right\rangle$, i.e.,

$$\boldsymbol{J}^2 \left|\pm\right\rangle = \bar{J} \left|\pm\right\rangle. \tag{3.45}$$

Similarly, making use of (3.43) we form

$$|\pm\rangle \equiv \boldsymbol{J}_\pm \left|\bar{J}J_3\right\rangle = \pm\left[\boldsymbol{J}_3, \boldsymbol{J}_\pm\right] \left|\bar{J}J_3\right\rangle = \pm \boldsymbol{J}_3 \boldsymbol{J}_\pm \left|\bar{J}J_3\right\rangle \mp \boldsymbol{J}_\pm \boldsymbol{J}_3 \left|\bar{J}J_3\right\rangle, \quad \text{i.e.,} \tag{3.46}$$

$$\boldsymbol{J}_3 \left|\pm\right\rangle = (J_3 \pm 1) \left|\pm\right\rangle. \tag{3.47}$$

That is to say, the state $|\pm\rangle \equiv \boldsymbol{J}_\pm \left|\bar{J}J_3\right\rangle$ is proportional to $\left|\bar{J}, J_3 + 1\right\rangle$. Therefore, $\boldsymbol{J}_+$ is named raising operator and $\boldsymbol{J}_-$ lowering operator.
The question arises, how often in succession the operator $\boldsymbol{J}_+$ can act on $\left|\bar{J}J_3\right\rangle$ or whether a positive integer k_{max} exists so that $\boldsymbol{J}_+^k \left|\bar{J}J_3\right\rangle = 0$ for $k > k_{\text{max}}$. We make up the inner product of two $|+\rangle$-states and take (1.45) and (3.42) into account: $0 \leq \langle + \mid + \rangle = \langle \boldsymbol{J}_+ (\bar{J}J_3) \mid \boldsymbol{J}_+ (\bar{J}J_3) \rangle = \left\langle \bar{J}J_3 \right| \boldsymbol{J}_- \boldsymbol{J}_+ \left|\bar{J}J_3\right\rangle$. For the operator product $\boldsymbol{J}_-\boldsymbol{J}_+$ we develop

$$\begin{aligned}
\boldsymbol{J}^2 &= \boldsymbol{J}_1^2 + \boldsymbol{J}_2^2 + \boldsymbol{J}_3^2 = \left(\frac{1}{2}(\boldsymbol{J}_+ + \boldsymbol{J}_-)\right)^2 + \left(\frac{1}{2\mathrm{i}}(\boldsymbol{J}_+ - \boldsymbol{J}_-)\right)^2 + \boldsymbol{J}_3^2 \\
&= \frac{1}{2}(\boldsymbol{J}_+\boldsymbol{J}_- + \boldsymbol{J}_-\boldsymbol{J}_+) + \boldsymbol{J}_3^2 = \frac{1}{2}(\boldsymbol{J}_+\boldsymbol{J}_- - \boldsymbol{J}_-\boldsymbol{J}_+ + 2\boldsymbol{J}_-\boldsymbol{J}_+) + \boldsymbol{J}_3^2 \\
&= \boldsymbol{J}_3 + \boldsymbol{J}_-\boldsymbol{J}_+ + \boldsymbol{J}_3^2\,.
\end{aligned} \tag{3.48}$$

Therefore, we have

$$\begin{aligned}
0 \leq \left\langle \bar{J}J_3 \right| \boldsymbol{J}_-\boldsymbol{J}_+ \left|\bar{J}J_3\right\rangle &= \left\langle \bar{J}J_3 \right| \boldsymbol{J}^2 - \boldsymbol{J}_3^2 - \boldsymbol{J}_3 \left|\bar{J}J_3\right\rangle \\
&= \left(\bar{J} - J_3(J_3+1)\right) \left\langle \bar{J}J_3 \mid \bar{J}J_3 \right\rangle = \bar{J} - J_3(J_3+1)\,.
\end{aligned} \tag{3.49}$$

That is why the quantity J_3, which increases by 1 with every application of $\boldsymbol{J}_+$, has to meet the following condition:

$$\bar{J} \geq J_3(J_3+1)\,, \tag{3.50}$$

and there must exist a maximal value $J_{3,\text{max}}$ which satisfies $\left|\bar{J}J_{3,\text{max}}\right\rangle \neq 0$ and $\boldsymbol{J}_+ \left|\bar{J}J_{3,\text{max}}\right\rangle = 0$. The inner product of this state with itself must vanish, i.e., $0 = \left\langle \bar{J}J_{3,\text{max}} \right| \boldsymbol{J}_-\boldsymbol{J}_+ \left|\bar{J}J_{3,\text{max}}\right\rangle = \left(\bar{J} - J_{3,\text{max}}(J_{3,\text{max}}+1)\right) \left\langle \bar{J}J_{3,\text{max}} \mid \bar{J}J_{3,\text{max}} \right\rangle$, which yields

$$\bar{J} = J_{3,\text{max}}(J_{3,\text{max}}+1)\,. \tag{3.51}$$

Analogously, one observes that the operator $\boldsymbol{J}_-$ cannot act arbitrarily often on the state $|\bar{J}J_3\rangle$. A minimal value $J_{3,\text{min}}$ exists and one proves for it:

$$\bar{J} = J_{3,\text{min}}\,(J_{3,\text{min}} - 1)\,. \tag{3.52}$$

Starting from an arbitrary state $|\bar{J}J_3\rangle$, the numbers k of raising steps and l falling steps are possible in the following sense:

$$\begin{array}{llll} \boldsymbol{J}_+^k\,|\bar{J}J_3\rangle \neq 0 & \text{but} & \boldsymbol{J}_+^{k+1}\,|\bar{J}J_3\rangle = 0 & \text{and} \\ \boldsymbol{J}_-^l\,|\bar{J}J_3\rangle \neq 0 & \text{but} & \boldsymbol{J}_-^{l+1}\,|\bar{J}J_3\rangle = 0\,. & \end{array} \tag{3.53}$$

Therefore, $J_{3,\text{max}} = J_3 + k$ and $J_{3,\text{min}} = J_3 - l$ hold, which we insert in (3.51) and (3.52) like this: $\bar{J} = (J_3 + k)\,(J_3 + k + 1) = (J_3 - l)\,(J_3 - l - 1)$. We rearrange this equation as follows: $(k + l + 1)\,2J_3 = (k + l + 1)\,(l - k)$ and obtain

$$J_3 = \frac{l-k}{2}\,, \quad \text{which yields} \tag{3.54}$$

$$J_{3,\text{max}} = J_3 + k = \frac{k+l}{2}\,, \quad J_{3,\text{min}} = J_3 - l = -\frac{k+l}{2} \text{ and} \tag{3.55}$$

$$\bar{J} = J_{3,\text{max}}\,(J_{3,\text{max}} + 1) = \frac{k+l}{2}\left(\frac{k+l}{2} + 1\right). \tag{3.56}$$

Thus, we have found a set of $k + l + 1$ eigenkets with the positive eigenvalue $\bar{J}$ of the operator $\boldsymbol{J}^2$ according to (3.56) and with the following J_3 values

$$\begin{array}{l} J_{3,\text{max}} = \frac{k+l}{2},\ \frac{k+l}{2} - 1\,, \ldots, J_3 + 1 = \frac{l-k}{2} + 1\,,\ J_3 = \frac{l-k}{2}\,, \\ J_3 - 1 = \frac{l-k}{2} - 1\,, \ldots, J_{3,\text{min}} = -\frac{k+l}{2}\,. \end{array} \tag{3.57}$$

As usual, we introduce the symbol $j = \frac{k+l}{2}$, and obtain for $\bar{J}$

$$\bar{J} = j\,(j+1)\,. \tag{3.58}$$

If $k + l$ is even, the quantity j is an integer, otherwise it is half integer. That is, j can take on the values $0,\ \frac{1}{2},\ 1,\ \frac{3}{2},\ \ldots$. The eigenvalue of $\boldsymbol{J}_3$ runs through the area between $J_{3,\text{max}} = j$ and $J_{3.\,\text{min}} = -j$ in steps amounting 1. From now on we name it m. This so-called magnetic quantum number is $j, j-1, \ldots, 0$ (or $\frac{1}{2},\ -\frac{1}{2}$)$, \ldots,$ $-j+1$, or $-j$. Of course m is only half integer if j is half integer.

Summing up, for an arbitrary function $|jm\rangle$ we make the statements

$$\boldsymbol{J}^2\,|jm\rangle = j\,(j+1)\,|jm\rangle\,, \quad j = 0,\ \frac{1}{2},\ 1,\ \frac{3}{2}, \ldots, \tag{3.59}$$

$$\boldsymbol{J}_3\,|jm\rangle = m\,|jm\rangle\,, \quad m = j,\ j-1, \ldots, -j+1,\ -j. \tag{3.60}$$

The states $|j\,j\rangle\,, |j, j-1\rangle\,, \ldots, |j, -j+1\rangle\,, |j, -j\rangle$ constitute a *function multiplet* (see Section 1.5) of the Lie algebra based on $\{\boldsymbol{J}_+, \boldsymbol{J}_-, \boldsymbol{J}_3\}$, (3.43), with the ordering number j. We name it $[j]$ or $M_{2j+1}\,(2j)$. From (3.47) we have deduced that $\boldsymbol{J}_+\,|jm\rangle$ is proportional to $|j, m+1\rangle$, that is,

$$\boldsymbol{J}_+\,|jm\rangle = N_{+,jm}\,|j, m+1\rangle\,. \tag{3.61}$$

We form the norm

$$\langle \boldsymbol{J}_+ (jm)| \boldsymbol{J}_+ |jm\rangle = N^*_{+,jm} \langle j, m+1| N_{+,jm} |j, m+1\rangle = |N_{+,jm}|^2 . \tag{3.62}$$

On the other hand we write with the help of (3.42) and (3.48):

$$\begin{aligned} \langle \boldsymbol{J}_+ (jm)| \boldsymbol{J}_+ |jm\rangle &= \langle jm| \boldsymbol{J}_+^\dagger \boldsymbol{J}_+ |jm\rangle = \langle jm| \boldsymbol{J}_- \boldsymbol{J}_+ |jm\rangle \\ &= \langle jm| \boldsymbol{J}^2 - \boldsymbol{J}_3 (\boldsymbol{J}_3 + 1) |jm\rangle = j(j+1) - m(m+1) . \end{aligned} \tag{3.63}$$

That is,

$$|N_{+,jm}|^2 = j(j+1) - m(m+1) . \tag{3.64}$$

Choosing the arbitrary phase factor +1 we write

$$\boldsymbol{J}_+ |jm\rangle = \sqrt{j(j+1) - m(m+1)}\, |j, m+1\rangle . \tag{3.65}$$

Analogously, one obtains

$$\boldsymbol{J}_- |jm\rangle = \sqrt{j(j+1) - m(m-1)}\, |j, m-1\rangle . \tag{3.66}$$

For the lowest $su(2)$-multiplets the following relations hold

$$\begin{aligned} \boldsymbol{J}_\pm \left| j = \frac{1}{2}, m = \mp\frac{1}{2} \right\rangle &= \left| \frac{1}{2}, \pm\frac{1}{2} \right\rangle , \\ \boldsymbol{J}_\pm |j = 1, m = \mp 1\rangle &= \sqrt{2}\, |1, 0\rangle , \\ \boldsymbol{J}_\pm |j = 1, m = 0\rangle &= \sqrt{2}\, |1, \pm 1\rangle , \\ \boldsymbol{J}_\pm \left| j = \frac{3}{2}, m = \mp\frac{3}{2} \right\rangle &= \sqrt{3} \left| \frac{3}{2}, \mp\frac{1}{2} \right\rangle , \\ \boldsymbol{J}_\pm \left| j = \frac{3}{2}, m = \mp\frac{1}{2} \right\rangle &= 2 \left| \frac{3}{2}, \pm\frac{1}{2} \right\rangle , \\ \boldsymbol{J}_\pm \left| j = \frac{3}{2}, m = \pm\frac{1}{2} \right\rangle &= \sqrt{3} \left| \frac{3}{2}, \pm\frac{3}{2} \right\rangle . \end{aligned} \tag{3.67}$$

In each equation (3.67) either the upper or the lower sign has to be taken everywhere. Other expressions with these operators and these states vanish.

The commutators (3.43) show that the operators $\boldsymbol{J}_+$, $\boldsymbol{J}_-$ and $\boldsymbol{J}_3$ constitute a Lie algebra. Due to (3.26) and (3.41) they read

$$\boldsymbol{J}_+ = \mathrm{i}(\boldsymbol{e}_1 + \mathrm{i}\boldsymbol{e}_2) , \quad \boldsymbol{J}_- = \mathrm{i}(\boldsymbol{e}_1 - \mathrm{i}\boldsymbol{e}_2) \quad \text{and} \quad \boldsymbol{J}_3 = \mathrm{i}\boldsymbol{e}_3 . \tag{3.68}$$

Thus, the elements of this algebra are complex linear combinations of $\mathbf{e}_1, \mathbf{e}_2$ and $\mathbf{e}_3$, i.e., they make up the *complexified version of* $su(2)$, which is named $sl(2, \mathbb{C})$ (see (3.7)). The inverse of (3.68) reads

$$\boldsymbol{e}_1 = \frac{1}{2\mathrm{i}} (\boldsymbol{J}_+ + \boldsymbol{J}_-) , \quad \boldsymbol{e}_2 = -\frac{1}{2} (\boldsymbol{J}_+ - \boldsymbol{J}_-) , \quad \boldsymbol{e}_3 = \frac{1}{\mathrm{i}} \boldsymbol{J}_3 , \tag{3.69}$$

which reveals that an operator $\boldsymbol{e}_i$ acting on a state $|j,m'\rangle$ generates a complex linear combination of such states $|j,m\rangle$, i.e., the result lies in the vector space over the field of complex numbers of the functions $\{|j,m\rangle\}$. Consequently, *the functions* $\{|j,m\rangle\}$ *are also associated to the Lie algebra* $su(2)$ and are named $su(2)$-multiplet. The representations $\left\{\underline{\underline{\Gamma}}^{(j)}(\boldsymbol{J}_i)\right\}$ or $\left\{\underline{\underline{\Gamma}}^{(j)}(\boldsymbol{e}_i)\right\}$ are irreducible (see Section 3.4).
The multiplet $\left|\frac{1}{2},\frac{1}{2}\right\rangle$, $\left|\frac{1}{2},-\frac{1}{2}\right\rangle$ is named fundamental $su(2)$-multiplet.

3.4 Irreducible representations of $su(2)$

According to (1.32), we can write the equations (3.65), (3.66) and (3.60) like this:

$$\boldsymbol{J}_i\,|jm\rangle = \sum_{m'=-j}^{j} \Gamma^{(j)}(\boldsymbol{J}_i)_{m'm}\,|jm'\rangle \quad \text{with } i=+,-,3. \tag{3.70}$$

Due to (1.36), the coefficients on the right hand side of (3.70) are matrix elements this way:

$$\Gamma^{(j)}(\boldsymbol{J}_i)_{m'm} = \langle jm'|\,\boldsymbol{J}_i\,|jm\rangle \quad \text{with } i=+,-,3. \tag{3.71}$$

The $su(2)$-multiplets $\{|jm\rangle\}$ are characterised by the quantum number j (see Section 3.3), that is why the $\Gamma's$ have a common index j. From Section 1.4 we know that the matrices $\underline{\underline{\Gamma}}^{(j)}(\boldsymbol{J}_i)$ are basis elements of a representation of the Lie algebra $\{\boldsymbol{J}_+,\boldsymbol{J}_-,\boldsymbol{J}_3\}$. In particular, due to (3.43) the following relations hold:

$$\begin{aligned}\left[\underline{\underline{\Gamma}}^{(j)}(\boldsymbol{J}_+),\underline{\underline{\Gamma}}^{(j)}(\boldsymbol{J}_-)\right] &= 2\underline{\underline{\Gamma}}^{(j)}(\boldsymbol{J}_3)\,,\\ \left[\underline{\underline{\Gamma}}^{(j)}(\boldsymbol{J}_3),\underline{\underline{\Gamma}}^{(j)}(\boldsymbol{J}_\pm)\right] &= \pm\underline{\underline{\Gamma}}^{(j)}(\boldsymbol{J}_\pm)\,.\end{aligned} \tag{3.72}$$

We calculate some matrix elements. Due to (3.67), for $j=\frac{1}{2}$ we have $\Gamma^{(\frac{1}{2})}(\boldsymbol{J}_+)_{\frac{1}{2}\frac{1}{2}} = \Gamma^{(\frac{1}{2})}(\boldsymbol{J}_+)_{-\frac{1}{2}\frac{1}{2}} = \Gamma^{(\frac{1}{2})}(\boldsymbol{J}_+)_{-\frac{1}{2},-\frac{1}{2}} = 0$ and $\Gamma^{(\frac{1}{2})}(\boldsymbol{J}_+)_{\frac{1}{2},-\frac{1}{2}} = 1$. For $j=\frac{1}{2}$, 1, and $\frac{3}{2}$ the matrices $\underline{\underline{\Gamma}}^{(j)}(\boldsymbol{J}_i)$ read like this:

$$\begin{aligned}
\underline{\underline{\Gamma}}^{(\frac{1}{2})}(\boldsymbol{J}_+) &= \begin{pmatrix} 0 & 1\\ 0 & 0\end{pmatrix},\ \underline{\underline{\Gamma}}^{(\frac{1}{2})}(\boldsymbol{J}_-) = \begin{pmatrix} 0 & 0\\ 1 & 0\end{pmatrix},\\
\underline{\underline{\Gamma}}^{(\frac{1}{2})}(\boldsymbol{J}_3) &= \begin{pmatrix} \frac{1}{2} & 0\\ 0 & -\frac{1}{2}\end{pmatrix},\\
\underline{\underline{\Gamma}}^{(1)}(\boldsymbol{J}_+) &= \begin{pmatrix} 0 & \sqrt{2} & 0\\ 0 & 0 & \sqrt{2}\\ 0 & 0 & 0\end{pmatrix},\ \underline{\underline{\Gamma}}^{(1)}(\boldsymbol{J}_-) = \begin{pmatrix} 0 & 0 & 0\\ \sqrt{2} & 0 & 0\\ 0 & \sqrt{2} & 0\end{pmatrix},\\
\underline{\underline{\Gamma}}^{(1)}(\boldsymbol{J}_3) &= \begin{pmatrix} 1 & 0 & 0\\ 0 & 0 & 0\\ 0 & 0 & -1\end{pmatrix},
\end{aligned} \tag{3.73}$$

$$\underline{\underline{\Gamma}}^{(\frac{3}{2})}(\boldsymbol{J}_+) = \begin{pmatrix} 0 & \sqrt{3} & 0 & 0 \\ 0 & 0 & 2 & 0 \\ 0 & 0 & 0 & \sqrt{3} \\ 0 & 0 & 0 & 0 \end{pmatrix}, \quad \underline{\underline{\Gamma}}^{(\frac{3}{2})}(\boldsymbol{J}_-) = \begin{pmatrix} 0 & 0 & 0 & 0 \\ \sqrt{3} & 0 & 0 & 0 \\ 0 & 2 & 0 & 0 \\ 0 & 0 & \sqrt{3} & 0 \end{pmatrix},$$

$$\underline{\underline{\Gamma}}^{(\frac{3}{2})}(\boldsymbol{J}_3) = \begin{pmatrix} \frac{3}{2} & 0 & 0 & 0 \\ 0 & \frac{1}{2} & 0 & 0 \\ 0 & 0 & -\frac{1}{2} & 0 \\ 0 & 0 & 0 & -\frac{3}{2} \end{pmatrix}. \tag{3.74}$$

Elementary calculations confirm that the matrices (3.73) and (3.74) satisfy the relations (3.72). Due to (3.69), the relations

$$\begin{aligned} \Gamma^{(j)}(\boldsymbol{e}_1) &= \frac{1}{2\mathrm{i}}\left(\Gamma^{(j)}(\boldsymbol{J}_+) + \Gamma^{(j)}(\boldsymbol{J}_-)\right), \\ \Gamma^{(j)}(\boldsymbol{e}_2) &= -\frac{1}{2}\left(\Gamma^{(j)}(\boldsymbol{J}_+) - \Gamma^{(j)}(\boldsymbol{J}_-)\right), \\ \Gamma^{(j)}(\boldsymbol{e}_3) &= \frac{1}{\mathrm{i}}\Gamma^{(j)}(\boldsymbol{J}_3) \end{aligned} \tag{3.75}$$

hold. Thus, we have

$$\begin{aligned} \underline{\underline{\Gamma}}^{(\frac{1}{2})}(\boldsymbol{e}_1) &= \frac{1}{2\mathrm{i}}\begin{pmatrix} 0 & 1 \\ 1 & 0 \end{pmatrix}, \quad \underline{\underline{\Gamma}}^{(\frac{1}{2})}(\boldsymbol{e}_2) = \frac{1}{2}\begin{pmatrix} 0 & -1 \\ 1 & 0 \end{pmatrix}, \\ \underline{\underline{\Gamma}}^{(\frac{1}{2})}(\boldsymbol{e}_3) &= \frac{1}{\mathrm{i}}\begin{pmatrix} \frac{1}{2} & 0 \\ 0 & -\frac{1}{2} \end{pmatrix}, \end{aligned} \tag{3.76}$$

in accordance with (3.1). The higher representations read

$$\begin{aligned} \underline{\underline{\Gamma}}^{(1)}(\boldsymbol{e}_1) &= \frac{1}{2\mathrm{i}}\begin{pmatrix} 0 & \sqrt{2} & 0 \\ \sqrt{2} & 0 & \sqrt{2} \\ 0 & \sqrt{2} & 0 \end{pmatrix}, \quad \underline{\underline{\Gamma}}^{(1)}(\boldsymbol{e}_2) = \frac{1}{2}\begin{pmatrix} 0 & -\sqrt{2} & 0 \\ \sqrt{2} & 0 & -\sqrt{2} \\ 0 & \sqrt{2} & 0 \end{pmatrix}, \\ \underline{\underline{\Gamma}}^{(1)}(\boldsymbol{e}_3) &= \frac{1}{\mathrm{i}}\begin{pmatrix} 1 & 0 & 0 \\ 0 & 0 & 0 \\ 0 & 0 & -1 \end{pmatrix}, \end{aligned} \tag{3.77}$$

$$\underline{\underline{\Gamma}}^{(\frac{3}{2})}(\boldsymbol{e}_1) = \frac{1}{2\mathrm{i}}\begin{pmatrix} 0 & \sqrt{3} & 0 & 0 \\ \sqrt{3} & 0 & 2 & 0 \\ 0 & 2 & 0 & \sqrt{3} \\ 0 & 0 & \sqrt{3} & 0 \end{pmatrix},$$

$$\underline{\underline{\Gamma}}^{(\frac{3}{2})}(\mathbf{e}_2) = \frac{1}{2}\begin{pmatrix} 0 & -\sqrt{3} & 0 & 0 \\ \sqrt{3} & 0 & -2 & 0 \\ 0 & 2 & 0 & -\sqrt{3} \\ 0 & 0 & \sqrt{3} & 0 \end{pmatrix},$$

$$\underline{\underline{\Gamma}}^{(\frac{3}{2})}(\mathbf{e}_3) = \frac{1}{\mathrm{i}}\begin{pmatrix} \frac{3}{2} & 0 & 0 & 0 \\ 0 & \frac{1}{2} & 0 & 0 \\ 0 & 0 & -\frac{1}{2} & 0 \\ 0 & 0 & 0 & -\frac{3}{2} \end{pmatrix}. \tag{3.78}$$

3.5 Direct products of irreducible representations and their function sets

For nuclear physics and the physics of elementary particles direct products of irreducible representations, their multiplets and their reduction are very important.

In this and in the next section we will deal with the Lie algebra $\{\boldsymbol{J}_+, \boldsymbol{J}_-, \boldsymbol{J}_3\}$ instead of the $su(2)$-algebra $\{\mathbf{e}_1, \mathbf{e}_2, \mathbf{e}_3\}$. The transformation relations (3.69) between both algebras and the corresponding equations (3.75) for the representations are simple. The multiplets $\{|jm\rangle\}$ are affiliated to both algebras.

Suppose, a square matrix $\underline{\underline{A}}$ with dimension d and a square matrix $\underline{\underline{B}}$ with dimension d' are given. Their **direct product** $\underline{\underline{A}} \otimes \underline{\underline{B}}$ is a matrix with the dimension $d \cdot d'$, the rows and the columns of which are denoted by pairs of indices. It reads

$$\left(\underline{\underline{A}} \otimes \underline{\underline{B}}\right)_{hs,kt} = A_{hk} \cdot B_{st}. \tag{3.79}$$

In the double index hs of the rows the quantity h grows slowly and s runs through the area of d' over and over. The double index kt of the columns behaves correspondingly. Equation (3.79) can be understood like this: replacing in $\underline{\underline{A}}$ every element A_{ij} by the matrix $A_{ij} \cdot \underline{\underline{B}}$ results in the direct product $\underline{\underline{A}} \otimes \underline{\underline{B}}$.

For the Lie algebra $\{\boldsymbol{J}_+, \boldsymbol{J}_-, \boldsymbol{J}_3\}$ the direct product $\underline{\underline{\Gamma}}^{(j)}(\boldsymbol{J}_i) \otimes \underline{\underline{1}}_{2j'+1}$ of the irreducible representation matrix $\underline{\underline{\Gamma}}^{(j)}(\boldsymbol{J}_i)$ (with dimensions $2j+1$) and of the identity matrix $\underline{\underline{1}}_{2j'+1}$ (with dimension $2j'+1$) is important. A general matrix element is written like this:

$$\left(\underline{\underline{\Gamma}}^{(j)}(\boldsymbol{J}_i) \otimes \underline{\underline{1}}_{2j'+1}\right)_{hs,kt} = \underline{\underline{\Gamma}}^{(j)}(\boldsymbol{J}_i)_{hk} \cdot \underline{\underline{1}}_{2j'+1,\ st}. \tag{3.80}$$

The matrix which is described by (3.80), reads as follows (we drop the arguments (J_i)):

$$\underline{\underline{\Gamma}}^{(j)} \otimes \underline{\underline{1}}_{2j'+1} \tag{3.81}$$

$$= \left(\begin{array}{ccccccccccc}
\Gamma_{11} & 0 & 0 & 0 & \Gamma_{12} & 0 & 0 & 0 & \Gamma_{13} & & \cdot \\
0 & \Gamma_{11} & 0 & 0 & 0 & \Gamma_{12} & 0 & 0 & & \cdot & \cdot \\
0 & 0 & \cdot & 0 & 0 & 0 & \cdot & 0 & & \cdot & \cdot \\
0 & 0 & 0 & \Gamma_{11} & 0 & 0 & 0 & \Gamma_{12} & & & \\
\Gamma_{21} & 0 & 0 & 0 & \Gamma_{22} & 0 & 0 & 0 & & & \\
0 & \Gamma_{21} & 0 & 0 & 0 & \Gamma_{22} & 0 & 0 & & & \\
0 & 0 & \cdot & 0 & 0 & 0 & \cdot & 0 & & & \\
0 & 0 & 0 & \Gamma_{21} & 0 & 0 & 0 & \Gamma_{22} & & & \\
\Gamma_{31} & & & & \cdot & & & & & & \cdot \\
& \cdot & & & & \cdot & & & & & \\
& & \cdot & & & & \cdot & & & & \\
& & & & & & & & & &
\end{array} \right)$$

The analogously defined direct product matrix $\underline{\underline{1}}_{2j+1} \otimes \underline{\underline{\Gamma}}^{(j')} (\boldsymbol{J}_i)$ reads as follows:

$$\underline{\underline{1}}_{2j'+1} \otimes \underline{\underline{\Gamma}}^{(j)} \tag{3.82}$$

$$= \left(\begin{array}{ccccccccccc}
\Gamma_{11} & \Gamma_{12} & \Gamma_{13} & \cdot & 0 & 0 & 0 & 0 & & & \\
\Gamma_{21} & \Gamma_{22} & & & 0 & 0 & 0 & 0 & & 0 & 0 \\
\Gamma_{31} & & \cdot & & 0 & 0 & 0 & 0 & & & \\
\cdot & & & \cdot & 0 & 0 & 0 & 0 & & & \\
0 & 0 & 0 & 0 & \Gamma_{11} & \Gamma_{12} & \Gamma_{13} & \cdot & & & \\
0 & 0 & 0 & 0 & \Gamma_{21} & \Gamma_{22} & & & & 0 & \\
0 & 0 & 0 & 0 & \Gamma_{31} & & \cdot & & & & \\
0 & 0 & 0 & 0 & \cdot & & & \cdot & & & \\
& & & & & & & & \Gamma_{11} & & \cdot \\
& 0 & & & & 0 & & & & \cdot & \\
& & & & & & & & & & \cdot \\
& & & & & & & & & & \\
& 0 & & & & & & & & &
\end{array} \right)$$

Now we show how two **representations** can be coupled to a **"direct product"** using the expressions above. We start with two function sets $\{|j, m\rangle\}$ and $\{|j', m'\rangle\}$ and form the set of all kinds of products $|jm\rangle \cdot |j'm'\rangle$. These quantities constitute a

$(2j+1)(2j'+1)$-dimensional vector space. Taking into account the relation between a real Lie algebra and the corresponding Lie group (see Section 2.2) and using matrix exponential functions, one shows (Cornwell, 1990, p. 412):

$$\boldsymbol{e}_i\left(|jm\rangle \cdot |j'm'\rangle\right) = \left(\boldsymbol{e}_i\,|jm\rangle\right) \cdot |j'm'\rangle + |jm\rangle \cdot \left(\boldsymbol{e}_i\,|j'm'\rangle\right), \tag{3.83}$$

which is given here without proof and which holds correspondingly for the most real Lie algebras. We put together the following linear combination:

$$\mathrm{i}(\boldsymbol{e}_1 + \mathrm{i}\boldsymbol{e}_2)\left(|jm\rangle \cdot |j'm'\rangle\right) = \left(\mathrm{i}\left(\boldsymbol{e}_1 + \mathrm{i}\boldsymbol{e}_2\right)|jm\rangle\right) \cdot |j'm'\rangle + |jm\rangle \cdot \left(\mathrm{i}\left(\boldsymbol{e}_1 + \mathrm{i}\boldsymbol{e}_2\right)|j'm'\rangle\right), \tag{3.84}$$

which stands for (see (3.68))

$$\boldsymbol{J}_+\left(|jm\rangle \cdot |j'm'\rangle\right) = \left(\boldsymbol{J}_+\,|jm\rangle\right) \cdot |j'm'\rangle + |jm\rangle \cdot \left(\boldsymbol{J}_+\,|j'm'\rangle\right). \tag{3.85}$$

Analogous relations hold for every $\boldsymbol{J}_i \quad (i = +, -, 3)$. We use the identity $|jm\rangle \equiv \sum_n \left(1_{(2j+1)}\right)_{nm} |jn\rangle$, insert (3.70) in the equation above and obtain

$$\begin{aligned} &\boldsymbol{J}_i\left(|jm\rangle \cdot |j'm'\rangle\right) \\ &= \sum_{nn'} \left(\Gamma^j\left(\boldsymbol{J}_i\right)_{nm}\left(1_{(2j'+1)}\right)_{n'm'} + \left(1_{(2j+1)}\right)_{nm} \Gamma^{j'}\left(\boldsymbol{J}_i\right)_{n'm'}\right)\left(|jn\rangle \cdot |j'n'\rangle\right). \end{aligned} \tag{3.86}$$

The resulting sum of two direct products on the right hand side is named

$$\underline{\underline{\Delta}}^{jj'}\left(\boldsymbol{J}_i\right) \equiv \underline{\underline{\Gamma}}^j\left(\boldsymbol{J}_i\right) \otimes \underline{\underline{1}}_{(2j'+1)} + \underline{\underline{1}}_{(2j+1)} \otimes \underline{\underline{\Gamma}}^{j'}\left(\boldsymbol{J}_i\right), \quad i = +, -, 3. \tag{3.87}$$

Thus, we can write

$$\boldsymbol{J}_i\left(|jm\rangle \cdot |j'm'\rangle\right) = \sum_{nn'} \left(\Delta^{jj'}\left(\boldsymbol{J}_i\right)\right)_{nn',mm'} \left(|jn\rangle \cdot |j'n'\rangle\right). \tag{3.88}$$

Although the matrices $\underline{\underline{\Delta}}^{jj'}\left(\boldsymbol{J}_i\right)$ are not simple direct products but a sum of such products (see(3.87)), they are frequently named "direct products" of $\underline{\underline{\Gamma}}^{(j)}\left(\boldsymbol{J}_i\right)$ and $\underline{\underline{\Gamma}}^{(j')}\left(\boldsymbol{J}_i\right)$. According to Section 1.4, the matrices $\underline{\underline{\Delta}}^{jj'}\left(\boldsymbol{J}_i\right)$ are representations of the Lie algebra $\{\boldsymbol{J}_+, \boldsymbol{J}_-, \boldsymbol{J}_3\}$, and the set $\{|jm\rangle \cdot |j'm'\rangle\}$ contains the affiliated functions.

The set of states $|jm\rangle \cdot |j'm'\rangle$ is marked by $[j] \otimes [j']$. It is named also tensor product or direct product of the multiplets $[j]$ and $[j']$.

We look at **two examples of "direct products" of two irreducible representations**. With the aid of (3.73) and (3.87) for $j = j' = \frac{1}{2}$ we obtain

$$\underline{\underline{\Delta}}^{\frac{1}{2}\frac{1}{2}}(\boldsymbol{J}_+) = \begin{pmatrix} 0 & 1 & 1 & 0 \\ 0 & 0 & 0 & 1 \\ 0 & 0 & 0 & 1 \\ 0 & 0 & 0 & 0 \end{pmatrix}, \quad \underline{\underline{\Delta}}^{\frac{1}{2}\frac{1}{2}}(\boldsymbol{J}_-) = \begin{pmatrix} 0 & 0 & 0 & 0 \\ 1 & 0 & 0 & 0 \\ 1 & 0 & 0 & 0 \\ 0 & 1 & 1 & 0 \end{pmatrix},$$

$$\underline{\underline{\Delta}}^{\frac{1}{2}\frac{1}{2}}(\boldsymbol{J}_3) = \begin{pmatrix} 1 & 0 & 0 & 0 \\ 0 & 0 & 0 & 0 \\ 0 & 0 & 0 & 0 \\ 0 & 0 & 0 & -1 \end{pmatrix}. \tag{3.89}$$

With elementary calculations one can check that the matrices $\underline{\underline{\Delta}}^{jj'}(\boldsymbol{J}_i)$ of (3.89) meet, in analogy to (3.43) and (3.72), the relations

$$\left[\underline{\underline{\Delta}}^{\frac{1}{2}\frac{1}{2}}(\boldsymbol{J}_+), \underline{\underline{\Delta}}^{\frac{1}{2}\frac{1}{2}}(\boldsymbol{J}_-)\right] = 2\underline{\underline{\Delta}}^{\frac{1}{2}\frac{1}{2}}(\boldsymbol{J}_3), \quad \left[\underline{\underline{\Delta}}^{\frac{1}{2}\frac{1}{2}}(\boldsymbol{J}_3), \underline{\underline{\Delta}}^{\frac{1}{2}\frac{1}{2}}(\boldsymbol{J}_\pm)\right] = \pm\underline{\underline{\Delta}}^{\frac{1}{2}\frac{1}{2}}(\boldsymbol{J}_\pm). \tag{3.90}$$

In accordance with (3.88), we obtain with the help of (3.89) the following first component of the product of the row vector of states with the matrix $\underline{\underline{\Delta}}^{\frac{1}{2}\frac{1}{2}}(\boldsymbol{J}_-)$:

$$\begin{aligned} &\boldsymbol{J}_- \left|\frac{1}{2}\right\rangle \left|\frac{1}{2}\right\rangle' = \\ &\left(\left(\left|\frac{1}{2}\right\rangle \left|\frac{1}{2}\right\rangle', \left|\frac{1}{2}\right\rangle \left|-\frac{1}{2}\right\rangle', \left|-\frac{1}{2}\right\rangle \left|\frac{1}{2}\right\rangle', \left|-\frac{1}{2}\right\rangle \left|-\frac{1}{2}\right\rangle' \right) \begin{pmatrix} 0 & 0 & 0 & 0 \\ 1 & 0 & 0 & 0 \\ 1 & 0 & 0 & 0 \\ 0 & 1 & 1 & 0 \end{pmatrix} \right)_1 \\ &= 0 \cdot \left|\frac{1}{2}\right\rangle \left|\frac{1}{2}\right\rangle' + 1 \cdot \left|\frac{1}{2}\right\rangle \left|-\frac{1}{2}\right\rangle' + 1 \left|-\frac{1}{2}\right\rangle \left|\frac{1}{2}\right\rangle' + 0 \cdot \left|-\frac{1}{2}\right\rangle \left|-\frac{1}{2}\right\rangle'. \end{aligned} \tag{3.91}$$

Only the m-values $1/2$ or $-1/2$ are written.

In the **second example** we combine multiplets with $j = 1$ and $j = 1/2$ (see (3.73) and (3.87)):

$$\underline{\underline{\Delta}}^{1,1/2}(\boldsymbol{J}_+) = \begin{pmatrix} 0 & 1 & \sqrt{2} & 0 & 0 & 0 \\ 0 & 0 & 0 & \sqrt{2} & 0 & 0 \\ 0 & 0 & 0 & 1 & \sqrt{2} & 0 \\ 0 & 0 & 0 & 0 & 0 & \sqrt{2} \\ 0 & 0 & 0 & 0 & 0 & 1 \\ 0 & 0 & 0 & 0 & 0 & 0 \end{pmatrix},$$

$$\underline{\underline{\Delta}}^{1,1/2}(\boldsymbol{J}_-) = \begin{pmatrix} 0 & 0 & 0 & 0 & 0 & 0 \\ 1 & 0 & 0 & 0 & 0 & 0 \\ \sqrt{2} & 0 & 0 & 0 & 0 & 0 \\ 0 & \sqrt{2} & 1 & 0 & 0 & 0 \\ 0 & 0 & \sqrt{2} & 0 & 0 & 0 \\ 0 & 0 & 0 & \sqrt{2} & 1 & 0 \end{pmatrix}, \tag{3.92}$$

$$\underline{\underline{\Delta}}^{1,1/2}(\boldsymbol{J}_3) = \begin{pmatrix} \frac{3}{2} & 0 & 0 & 0 & 0 & 0 \\ 0 & \frac{1}{2} & 0 & 0 & 0 & 0 \\ 0 & 0 & \frac{1}{2} & 0 & 0 & 0 \\ 0 & 0 & 0 & -\frac{1}{2} & 0 & 0 \\ 0 & 0 & 0 & 0 & -\frac{1}{2} & 0 \\ 0 & 0 & 0 & 0 & 0 & -\frac{3}{2} \end{pmatrix}.$$

As in the first example one checks that the $\underline{\underline{\Delta}}^{1,1/2}(\boldsymbol{J}_i)$'s are basis matrices of the representation with affiliated functions $|1\ n\rangle \left|\frac{1}{2}n'\right\rangle$ $(n = 1, 0, -1;\ n' = 1/2, -1/2)$.

3.6 Reduction of direct products of $su(2)$-representations and multiplets

In connection with (3.88), we noted that the matrices $\underline{\underline{\Delta}}^{jj'}(\boldsymbol{J}_i)$ are a representation of the algebra $\{\boldsymbol{J}_+, \boldsymbol{J}_-, \boldsymbol{J}_3\}$. Although the matrices in (3.89) and (3.92) have no block diagonal form, they are not irreducible, which holds generally. Making use of a similarity transformation as in Section 1.5, one can bring the matrices $\underline{\underline{\Delta}}^{jj'}(\boldsymbol{J}_i)$ in a block diagonal form, which reveals the set of irreducible representations to which the representation can be reduced. At the same time, one obtains the multiplets of the involved irreducible representations.

We start with the **transformation of the function products** $(|jm\rangle \cdot |j'm'\rangle)$. The theory of direct products of irreducible representations shows that there exist non-singular matrices $\underline{\underline{S}}$ which transform the function product according to (1.60) in such a way that the resulting functions have again the properties of $su\,(2)$-functions with the parameters M (eigenvalue of $\boldsymbol{J}_3$), and J $\;(J\,(J+1)$ is eigenvalue of $\boldsymbol{J}^2)$. Thus, on can write

$$|JM\rangle = \sum_{mm'} S^{jj'}_{mm',JM}\left(|jm\rangle \cdot |j'm'\rangle\right). \tag{3.93}$$

The functions $|JM\rangle$ obey also the step operators $\boldsymbol{J}_+$ and $\boldsymbol{J}_-$. The coefficients $S^{jj'}_{mm',JM}$ are named *Clebsch–Gordan coefficients* (CGc) and they are frequently marked by $(jmj'm'|JM)$. They vanish if $m + m' \neq M$ and if $j + j' > J$ or $|j - j'| < J$, that is, if the triangle conditions are violated. The values of J are $j+j', j+j'-1, \ldots, |j-j'|+1$, or $|j-j'|$. Thus, we give without proof:

$$S^{jj'}_{mm',JM} = (jmj'm'|JM)\,. \tag{3.94}$$

In the literature the symmetry properties of the Clebsch–Gordan coefficients are made visible especially by means of the 3-j symbol $\begin{pmatrix} j & j' & J \\ m & m' & -M \end{pmatrix}$, which is defined as follows:

$$(jmj'm'|JM) = (-1)^{j-j'+M}\sqrt{2J+1}\begin{pmatrix} j & j' & J \\ m & m' & -M \end{pmatrix}. \tag{3.95}$$

Edmonds (1996) gave algebraic expressions for some 3-j symbols. Numerical tables were published by Rotenberg, Bivins et. al. (1959) and Brussaard and Glaudemans (1977).

If a parameter of a CGc is an algebraic expression, it is usual to fence it in by commas. For $j' = 1/2$ the Clebsch–Gordan coefficients $(j, M-m', j' = \frac{1}{2}, m'|JM)$ are given (without proof) by the following expressions:

$$\begin{array}{c|cc} & m' = \frac{1}{2} & m' = -\frac{1}{2} \\ \hline J = j + \frac{1}{2} & \left(\frac{j+M+\frac{1}{2}}{2j+1}\right)^{\frac{1}{2}} & \left(\frac{j-M+\frac{1}{2}}{2j+1}\right)^{\frac{1}{2}} \\ J = j - \frac{1}{2} & -\left(\frac{j-M+\frac{1}{2}}{2j+1}\right)^{\frac{1}{2}} & \left(\frac{j+M+\frac{1}{2}}{2j+1}\right)^{\frac{1}{2}} \end{array} \quad \text{with } j' = \frac{1}{2}. \tag{3.96}$$

With the help of (3.94) and (3.96) we calculate, for example, the matrix element $S^{j=j'=\frac{1}{2}}_{m=-\frac{1}{2},m'=\frac{1}{2},J=1,M=0} = (\frac{1}{2}, -\frac{1}{2}, \frac{1}{2}\ \frac{1}{2}|10) = \left(\frac{\frac{1}{2}+0+\frac{1}{2}}{1+1}\right)^{\frac{1}{2}} = \frac{1}{\sqrt{2}}$, and we obtain finally the following transformation matrix $\underline{\underline{S}}^{jj'}$ for $j = j' = 1/2$:

$$\underline{\underline{S}}^{\frac{1}{2}\frac{1}{2}} = \begin{pmatrix} 1 & 0 & 0 & 0 \\ 0 & \frac{1}{\sqrt{2}} & 0 & \frac{1}{\sqrt{2}} \\ 0 & \frac{1}{\sqrt{2}} & 0 & -\frac{1}{\sqrt{2}} \\ 0 & 0 & 1 & 0 \end{pmatrix}, \tag{3.97}$$

where the columns are indicated by $(J, M) =$ (1,1), (1,0) , (1,–1), (0,0) and the rows by $(m, m') =$ (1/2,1/2), (1/2,–1/2), (–1/2,1/2), (–1/2,–1/2). For $j = 1$

and $j' = 1/2$ we have

$$\underline{\underline{S}}^{1,1/2} = \begin{pmatrix} 1 & 0 & 0 & 0 & 0 & 0 \\ 0 & \frac{1}{\sqrt{3}} & 0 & 0 & \sqrt{\frac{2}{3}} & 0 \\ 0 & \sqrt{\frac{2}{3}} & 0 & 0 & -\frac{1}{\sqrt{3}} & 0 \\ 0 & 0 & \sqrt{\frac{2}{3}} & 0 & 0 & \frac{1}{\sqrt{3}} \\ 0 & 0 & \frac{1}{\sqrt{3}} & 0 & 0 & -\sqrt{\frac{2}{3}} \\ 0 & 0 & 0 & 1 & 0 & 0 \end{pmatrix}. \tag{3.98}$$

We now work on the **"direct product" of representations** applied in (3.88), i.e., in $\boldsymbol{J}_i\left(|jm\rangle \cdot |j'm'\rangle\right) = \sum\limits_{nn'} \left(\Delta^{jj'}\left(\boldsymbol{J}_i\right)\right)_{nn',mm'} \left(|jn\rangle \cdot |j'n'\rangle\right)$ for $i = +, -, 3$. We multiply the equation by $S^{jj'}_{mm',JM}$, sum over m and m' and obtain

$$\begin{aligned} &\boldsymbol{J}_i \sum_{mm'} S^{jj'}_{mm',JM} \left(|jm\rangle \cdot |j'm'\rangle\right) \\ &= \sum_{nn'} \left(\sum_{mm'} \left(\Delta^{jj'}\left(\boldsymbol{J}_i\right)\right)_{nn',mm'} S^{jj}_{mm',JM} \right) \left(|jn\rangle \cdot |j'n'\rangle\right). \end{aligned} \tag{3.99}$$

The sum on the left hand side represents $|JM\rangle$, (3.93). On the right hand side we make use of the identity $\left(|jn\rangle \cdot |j'n'\rangle\right) \equiv \sum\limits_{pp'} \delta_{pn}\delta_{p'n'} \left(|jp\rangle \cdot |j'p'\rangle\right)$ and of the expression of an element of the identity matrix

$$\sum_{J'M'} \left(S^{jj'}\right)^{-1}_{J'M',nn'} S^{jj'}_{pp',J'M'} = (1)_{pp',nn'} = \delta_{pn}\delta_{p'n'},$$

and obtain

$$\begin{aligned} \boldsymbol{J}_i \, |JM\rangle = \quad & \sum_{J'M'} \sum_{nn'} \left(S^{jj'}\right)^{-1}_{J'M'nn'} \left(\sum_{mm'} \left(\Delta^{jj'}\left(\boldsymbol{J}_i\right)\right)_{nn'mm'} S^{jj'}_{mm'JM} \right) \\ & \cdot \sum_{pp'} S^{jj'}_{pp'J'M'} \left(|jp\rangle \cdot |j'p'\rangle\right). \end{aligned} \tag{3.100}$$

Using (3.93) again, we can write

$$\boldsymbol{J}_i \, |JM\rangle = \sum_{J'M'} \left(\left(\underline{\underline{S}}^{jj'}\right)^{-1} \underline{\underline{\Delta}}^{jj'}\left(\boldsymbol{J}_i\right) \underline{\underline{S}}^{jj'} \right)_{J'M',JM} |J'M'\rangle. \tag{3.101}$$

Due to (3.60), (3.65) and (3.66), the operators $\boldsymbol{J}_i$ have no influence on J. Therefore, in (3.101) both sides must have the same eigenvalue of $\boldsymbol{J}^2$. That is, $J' = J$ holds, or all the matrix elements

$$\left(\left(\underline{\underline{S}}^{jj'}\right)^{-1} \underline{\underline{\Delta}}^{jj'}\left(\boldsymbol{J}_i\right) \underline{\underline{S}}^{jj'} \right)_{J'M',JM} \equiv \left(\underline{\underline{\Delta}}^{jj'}\left(\boldsymbol{J}_i\right)\right)'_{J'M',JM} \tag{3.102}$$

with $J' \neq J$ vanish. That is to say, $\left(\underline{\underline{\Delta}}^{jj'}(\boldsymbol{J}_i)\right)'$ has a block diagonal form, for instance, like this:

$$\left(\underline{\underline{\Delta}}^{jj'}(\boldsymbol{J}_i)\right)' = \begin{pmatrix} \cdot & \cdot & \cdot & & & & & \\ \cdot & \underline{\underline{\Gamma}}^{(J_a)}(\boldsymbol{J}_i) & \cdot & & 0 & & & 0 \\ \cdot & \cdot & \cdot & & & & & \\ & & & \cdot & \cdot & \cdot & & \\ & 0 & & \cdot & \underline{\underline{\Gamma}}^{(J_b)}(\boldsymbol{J}_i) & \cdot & & 0 \\ & & & \cdot & \cdot & \cdot & & \\ & & & & & & \cdot & \\ & 0 & & & 0 & & & \cdot \end{pmatrix}. \tag{3.103}$$

To $J = J_b$, say, belong $2J_b+1$ values M and the same number of values M', which constitute a square region in $\left(\underline{\underline{\Delta}}^{jj'}(\boldsymbol{J}_i)\right)'$. So it is evident that the representation $\left\{\underline{\underline{\Delta}}^{jj'}(\boldsymbol{J}_i)\right\}$ is reducible and the blocks appear in $\left(\underline{\underline{\Delta}}^{jj'}(\boldsymbol{J}_i)\right)'$, which are characterized by J_a, J_b, ... The relation (3.101) reduces to

$$\begin{aligned} \boldsymbol{J}_i\,|JM\rangle &= \sum_{M'}\left(\left(\underline{\underline{S}}^{jj'}\right)^{-1}\underline{\underline{\Delta}}^{jj'}\underline{\underline{S}}^{jj'}\right)_{JM',JM}|JM'\rangle \\ &\equiv \sum_{M'}\left(\underline{\underline{\Delta}}^{jj'}(\boldsymbol{J}_i)\right)'_{JM',JM}|JM'\rangle \qquad \text{for } i=+,-,3. \end{aligned} \tag{3.104}$$

Formally it is identical to (3.70) and describes the relations in a square block of (3.103), characterized by J, which is an irreducible representation with the multiplet $\{|JM\rangle\}$.

We now calculate numerically the matrices $\left(\underline{\underline{\Delta}}^{jj'}(\boldsymbol{J}_i)\right)'$, (3.102), for the examples (3.89) and (3.92). It is not difficult to find the inverse of the matrix $\underline{\underline{S}}^{jj'}$. Clebsch–Gordan coefficients meet the following orthogonality relation:

$$\sum_{m_a m_b}(j_a m_a j_b m_b|JM)\,(j_a m_a j_b m_b|J'M') = \delta_{JJ'}\delta_{MM'}. \tag{3.105}$$

The right hand side is the element $(JM, J'M')$ of the identity matrix $\underline{\underline{1}}$, where we use double indices (JM). Due to (3.94), the second CGc is identical to $S^{j_a j_b}_{m_a m_b, J'M'}$. If we denote the first CGc in (3.105) by $\left(S^{j_a j_b}\right)^{-1}_{JM, m_a m_b}$, we have

$$\begin{aligned} &\sum_{m_a m_b}\left(S^{j_a j_b}\right)^{-1}_{JM,m_a m_b} S^{j_a j_b}_{m_a m_b,J'M'} = 1_{(JM),(J'M')} \\ &\text{or}\quad \left(\underline{\underline{S}}^{j_a j_b}\right)^{-1}\cdot\underline{\underline{S}}^{j_a j_b} = \underline{\underline{1}}\,. \end{aligned} \tag{3.106}$$

We see that our denomination

$$\left(S^{j_a j_b}\right)^{-1}_{JM,m_a m_b} = (j_a m_a j_b m_a | JM) \tag{3.107}$$

is correct. Making use of (3.97), (3.107) and (3.89), elementary calculations yield

$$\left(\underline{\underline{\Delta}}^{\frac{1}{2}\frac{1}{2}}(\boldsymbol{J}_+)\right)' \equiv \left(\underline{\underline{S}}^{\frac{1}{2}\frac{1}{2}}\right)^{-1} \underline{\underline{\Delta}}^{\frac{1}{2}\frac{1}{2}}(\boldsymbol{J}_+)\,\underline{\underline{S}}^{\frac{1}{2}\frac{1}{2}} = \begin{pmatrix} 0 & \sqrt{2} & 0 & 0 \\ 0 & 0 & \sqrt{2} & 0 \\ 0 & 0 & 0 & 0 \\ 0 & 0 & 0 & 0 \end{pmatrix},$$

$$\left(\underline{\underline{\Delta}}^{\frac{1}{2}\frac{1}{2}}(\boldsymbol{J}_-)\right)' = \begin{pmatrix} 0 & 0 & 0 & 0 \\ \sqrt{2} & 0 & 0 & 0 \\ 0 & \sqrt{2} & 0 & 0 \\ 0 & 0 & 0 & 0 \end{pmatrix}, \quad \left(\underline{\underline{\Delta}}^{\frac{1}{2}\frac{1}{2}}(\boldsymbol{J}_3)\right)' = \begin{pmatrix} 1 & 0 & 0 & 0 \\ 0 & 0 & 0 & 0 \\ 0 & 0 & -1 & 0 \\ 0 & 0 & 0 & 0 \end{pmatrix}, \tag{3.108}$$

which we write by means of (3.73) like this:

$$\left(\underline{\underline{\Delta}}^{\frac{1}{2}\frac{1}{2}}(\boldsymbol{J}_i)\right)' = \begin{pmatrix} \cdot & \cdot & \cdot & 0 \\ \cdot & \underline{\underline{\Gamma}}^{(1)}(\boldsymbol{J}_i) & \cdot & 0 \\ \cdot & \cdot & \cdot & 0 \\ 0 & 0 & 0 & \underline{\underline{\Gamma}}^{(0)}(\boldsymbol{J}_i) \end{pmatrix}, \quad i = +, -, 3. \tag{3.109}$$

For the second example one obtains

$$\left(\underline{\underline{\Delta}}^{1,1/2}(\boldsymbol{J}_+)\right)' = \begin{pmatrix} 0 & \sqrt{3} & 0 & 0 & 0 & 0 \\ 0 & 0 & 2 & 0 & 0 & 0 \\ 0 & 0 & 0 & \sqrt{3} & 0 & 0 \\ 0 & 0 & 0 & 0 & 0 & 0 \\ 0 & 0 & 0 & 0 & 0 & 1 \\ 0 & 0 & 0 & 0 & 0 & 0 \end{pmatrix} \tag{3.110}$$

and can summarize

$$\left(\underline{\underline{\Delta}}^{1,1/2}(\boldsymbol{J}_i)\right)' = \begin{pmatrix} \cdot & \cdot & \cdot & \cdot & 0 & 0 \\ \cdot & \underline{\underline{\Gamma}}^{(\frac{3}{2})}(\boldsymbol{J}_i) & \cdot & \cdot & 0 & 0 \\ \cdot & \cdot & \cdot & \cdot & 0 & 0 \\ \cdot & \cdot & \cdot & \cdot & 0 & 0 \\ 0 & 0 & 0 & 0 & \underline{\underline{\Gamma}}^{(\frac{1}{2})}(\boldsymbol{J}_i) & \cdot \\ 0 & 0 & 0 & 0 & \cdot & \cdot \end{pmatrix}, \quad i = +, -, 3. \tag{3.111}$$

We write this equation symbolically like this:

$$\left(\underline{\underline{\Delta}}^{1,1/2}(\boldsymbol{J}_i)\right)' = \sum_{J=1/2}^{3/2} \oplus\, n_J^{1,1/2}\, \underline{\underline{\Gamma}}^{(J)}(\boldsymbol{J}_i) \quad \text{with } n_{1/2}^{1,1/2} = n_{3/2}^{1,1/2} = 1. \tag{3.112}$$

The numbers $n_J^{jj'}$ specify how often the irreducible representation $\underline{\underline{\Gamma}}^{(J)}(\boldsymbol{J}_i)$ appears. It can be shown that for $su(2)$ generally the relation

$$\begin{aligned}\left(\underline{\underline{\Delta}}^{jj'}(\boldsymbol{J}_i)\right)' &= \left(\left(\underline{\underline{S}}^{jj'}\right)^{-1}\underline{\underline{\Delta}}^{jj'}(\boldsymbol{J}_i)\underline{\underline{S}}^{jj'}\right) \\ &= \left(\underline{\underline{S}}^{jj'}\right)^{-1}\left(\underline{\underline{\Gamma}}^{(j)}(\boldsymbol{J}_i)\otimes\underline{\underline{1}}^{(j')}+\underline{\underline{1}}^{(j)}\otimes\underline{\underline{\Gamma}}^{(j')}(\boldsymbol{J}_i)\right)\underline{\underline{S}}^{jj'} \\ &= \underline{\underline{\Gamma}}^{j+j'}(\boldsymbol{J}_i)\oplus\underline{\underline{\Gamma}}^{j+j'-1}(\boldsymbol{J}_i)\oplus\ldots\oplus\underline{\underline{\Gamma}}^{|j-j'|+1}(\boldsymbol{J}_i)\oplus\underline{\underline{\Gamma}}^{|j-j'|}(\boldsymbol{J}_i)\end{aligned} \tag{3.113}$$

holds. That is, here we have $n_J^{jj''} = 1$, which we found already in (3.112) and (3.109).

The functions which belong to the representation (3.111) are calculated with the aid of (3.93) and (3.98) like this:

$$\begin{aligned}\left|\frac{3}{2},\frac{3}{2}\right\rangle &= S^{j=1,j'=1/2}_{m=1,m'=1/2,J'=M'=3/2}\,|j=1,m=1\rangle\left|j'=\frac{1}{2},m'=\frac{1}{2}\right\rangle \\ &= |1,1\rangle\left|\frac{1}{2}\frac{1}{2}\right\rangle, \\ \left|\frac{3}{2},\frac{1}{2}\right\rangle &= \frac{1}{\sqrt{2}}|1,1\rangle\left|\frac{1}{2},-\frac{1}{2}\right\rangle-\sqrt{\frac{2}{3}}|1,0\rangle\left|\frac{1}{2}\frac{1}{2}\right\rangle, \\ . &= . \\ . &= . \\ \left|\frac{1}{2},-\frac{1}{2}\right\rangle &= \frac{1}{\sqrt{3}}|1,0\rangle\left|\frac{1}{2},-\frac{1}{2}\right\rangle-\sqrt{\frac{3}{2}}|1,-1\rangle\left|\frac{1}{2}\frac{1}{2}\right\rangle.\end{aligned} \tag{3.114}$$

The set of states $\left|\frac{3}{2}\frac{3}{2}\right\rangle$, $\left|\frac{3}{2},\frac{1}{2}\right\rangle$, $\left|\frac{3}{2},-\frac{1}{2}\right\rangle$, $\left|\frac{3}{2},-\frac{3}{2}\right\rangle$, $\left|\frac{1}{2},\frac{1}{2}\right\rangle$, $\left|\frac{1}{2},-\frac{1}{2}\right\rangle$ is marked by

$$\left[\frac{3}{2}\right]\oplus\left[\frac{1}{2}\right], \tag{3.115}$$

where we made use of the symbol $[j]$ (see (3.60/61)). Here we use the symbol $\oplus$ for the combination of multiplets. Because the function set (3.115) results from the direct product of the irreducible representations $\left\{\underline{\underline{\Gamma}}^{(1)}(\boldsymbol{J}_i)\right\}$ and $\left\{\underline{\underline{\Gamma}}^{(\frac{1}{2})}(\boldsymbol{J}_i)\right\}$, one says that it is the direct product or tensor product or Kronecker product of the multiplets $[1]$ and $\left[\frac{1}{2}\right]$, and one writes

$$[1]\otimes\left[\frac{1}{2}\right]\triangleq\left[\frac{3}{2}\right]\oplus\left[\frac{1}{2}\right]. \tag{3.116}$$

From (3.109) we learn that

$$\left[\frac{1}{2}\right]\otimes\left[\frac{1}{2}\right]\triangleq[1]\oplus[0]. \tag{3.117}$$

Many authors use an equals sign instead of $\triangleq$ in (3.116), which is a questionable practice between two sets of states. Equation (3.116) says that the functions of the set on the right hand side can be calculated with the set on the left using the matrix $\underline{\underline{S}}^{jj'}$ according to (3.93). Since $\underline{\underline{S}}^{jj'}$ is non-singular the transformation can be inverted.

Due to (3.113), the general formula for the reduction of the direct product of two $su(2)$-multiplets reads as follows:

$$[j] \otimes [j'] \triangleq [j+j'] \oplus [j+j'-1] \oplus \ldots \oplus [|j-j'|+1] \oplus [|j-j'|] \,. \tag{3.118}$$

The right hand side of (3.118) is named *Clebsch–Gordan series.* In analogy to (3.113), in (3.118) every multiplet appears once ($n_J^{jj'} = 1$).

The following considerations support the statement that $n_J^{jj'} = 1$ for $su(2)$. Since the matrices in (3.112) have the same dimension as $\underline{\underline{\Delta}}^{jj'}(\boldsymbol{J}_i)$, (3.87), namely $(2j+1)(2j'+1)$, due to (3.112) the relation

$$(2j+1)(2j'+1) = \sum_J n_J^{jj'}(2J+1) \tag{3.119}$$

must hold. With $n_J^{jj'} = 1$ and with $j > j'$ the right hand side of (3.119) reads

$$\begin{aligned}
&\sum_{J=j-j'}^{j+j'} (2J+1) \\
&= 2\left[(j-j') + (j-j'+1) + \ldots + (j+j')\right] + j + j' - (j-j'-1) \\
&= 2\left[(j+j')(j+j'+1)/2 - (j-j'-1)(j-j')/2\right] + 2j' + 1 \\
&= (2j+1)(2j'+1) \,,
\end{aligned} \tag{3.120}$$

in accordance with the left hand side of (3.119).

3.7 Graphical reduction of direct products of $su(2)$-multiplets

The multiplet of the irreducible representation $\left\{\underline{\underline{\Gamma}}^{(j)}(\boldsymbol{J}_i)\right\}$, which comprises the states $|j\,j\rangle\,,\ |j, j-1\rangle\,, \ldots, |j,-j\rangle$ is denoted by $[j]$ or $M_{2j+1}(2j)$ (see (3.60/61)). We depict it by $2j+1$ points on a line like this:

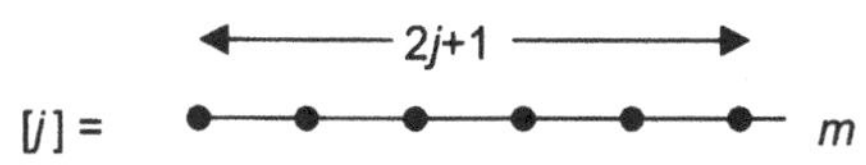

Figure 3.7.1. Graphical representation of the multiplet $[j]$.

The function set $[j]\otimes[j']$ of a direct product representation $\left\{\underline{\underline{\Delta}}^{jj'}(\boldsymbol{J}_i)\right\}$ comprises $(2j+1)(2j'+1)$ function products $|jm\rangle\,|j'm'\rangle$ (see (3.86)). We depict the set in such a way that we draw first $[j\ ]$ and then set the multiplet $[j']$ repeatedly so that its centre appears in every point of $[j\ ]$, which we delete afterwards. In this way all the values $M = m+m'$ of the set are outlined. For example, the multiplet product $[1]\otimes[3/2]$ is made up this way:

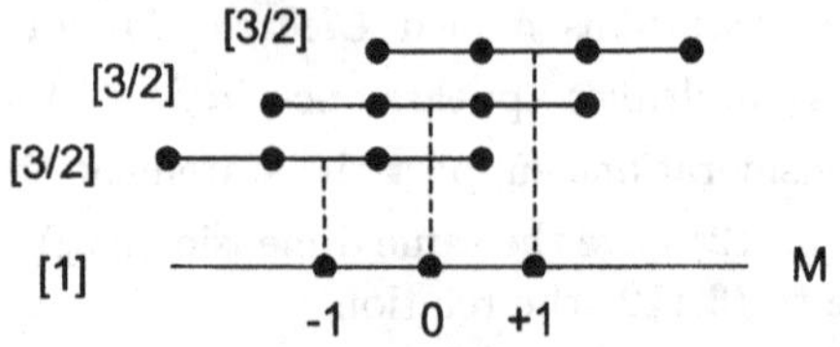

Figure 3.7.2. Direct product of the multiplets [1] and [3/2].

I.e., over every point of the [1]-triplet a [3/2]-quadruplet is placed. The value $M = 5/2$ appears once, $M = 3/2$ twice, etc. Preserving the number of points and the M-values, we rearrange the points in a symmetric way as multiplets on different levels. Instead of Figure 3.7.2, we obtain the following arrangement,

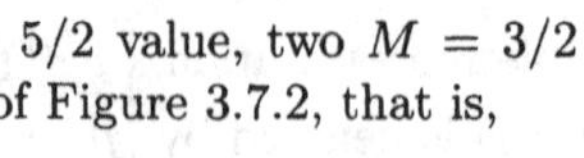

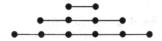

Figure 3.7.3. Direct sum of the multiplets [5/2], [3/2] and [1/2].

which has again one $M = 5/2$ value, two $M = 3/2$ values, etc. We say that it "equals" to the formation of Figure 3.7.2, that is,

$$[1]\otimes[3/2] \triangleq [5/2]\oplus[3/2]\oplus[1/2]\,, \qquad (3.121)$$

which coincides exactly with (3.118).

The next **example** deals with $[3]\otimes[1]$ in Figure 3.7.4.

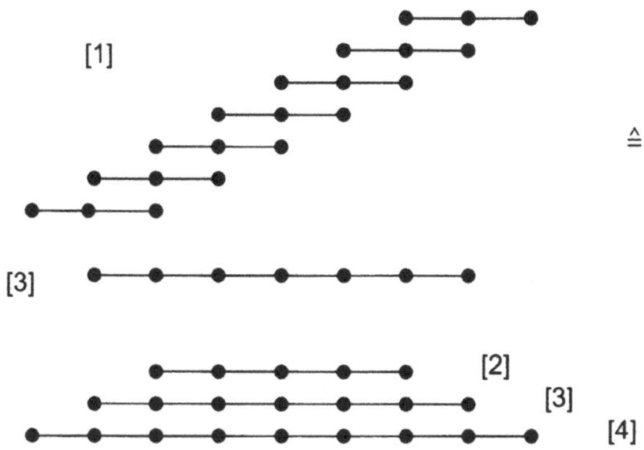

Figure 3.7.4. Direct product $[3] \otimes [1]$.

We read $[3] \otimes [1] \triangleq [4] \oplus [3] \oplus [2]$.

By this graphical method it is easy to reduce **multiple direct products** of multiplets. For example, in order to form $[1/2] \otimes [1/2] \otimes [1/2]$ we treat first $[1/2] \otimes [1/2]$ and then we attach $[1/2]$ as follows:

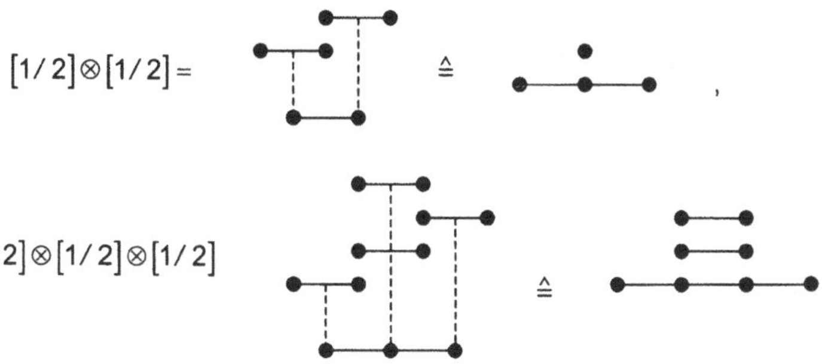

Figure 3.7.5. Direct product $[1/2] \otimes [1/2] \otimes [1/2]$.

i.e., we have the relation $[1/2] \otimes [1/2] \otimes [1/2] \triangleq [3/2] \oplus 2\,[1/2]$ for the fundamental multiplet of $su(2)$. In the standard model of particle physics multiple products of fundamental multiplets in $su(3)$ and in $su(4)$ are very important.

Chapter 4

The Lie algebra $su(3)$

The basis elements and the generators of the algebra $su(3)$ are given. The rank two of this algebra is derived from the table of the structure constants. The algebra $su(3)$ contains three subalgebras $su(2)$. Their complexified variants contain the "third" operators $\boldsymbol{T}_3, \boldsymbol{U}_3$ and $\boldsymbol{V}_3$, which characterise the states of the $su(3)$-multiplets. By means of the corresponding step operators, the structure of the multiplets and the pairs of their ordering numbers are shown. Specifying the individual states of a multiplet, their multiplicity and the dimension of the entire multiplet are found. The smallest $su(3)$-multiplets are given graphically.

The so-called hypercharge is introduced as an eigenvalue and interpreted graphically. Starting from a multiplet, an irreducible representation is calculated in detail. The quadratic Casimir operator is introduced and its eigenvalue is derived.

The graphical method for direct products of multiplets is taken from $su(2)$ and Clebsch–Gordan series are performed. The corresponding technique with Young diagrams is explained.

4.1 The generators of the $su(3)$-algebra

The anti-Hermitian 3×3-matrices with vanishing trace constitute the real Lie algebra $su(3)$ (see Section 2.2). Due to (2.15), we write the anti-Hermitian basis matrices as $\underline{\underline{e}}_i = -\frac{1}{2}\mathrm{i}\underline{\underline{\lambda}}_i$, where the generators $\underline{\underline{\lambda}}_i$ are Hermitian basis matrices. By means of (2.5), the matrices $\underline{\underline{\lambda}}_i$ of $su(3)$ can be written this way:

$$\underline{\underline{\lambda}}_1 = \begin{pmatrix} 0 & 1 & 0 \\ 1 & 0 & 0 \\ 0 & 0 & 0 \end{pmatrix}, \quad \underline{\underline{\lambda}}_2 = \begin{pmatrix} 0 & -\mathrm{i} & 0 \\ \mathrm{i} & 0 & 0 \\ 0 & 0 & 0 \end{pmatrix}, \quad \underline{\underline{\lambda}}_3 = \begin{pmatrix} 1 & 0 & 0 \\ 0 & -1 & 0 \\ 0 & 0 & 0 \end{pmatrix},$$
$$\underline{\underline{\lambda}}_4 = \begin{pmatrix} 0 & 0 & 1 \\ 0 & 0 & 0 \\ 1 & 0 & 0 \end{pmatrix}, \quad \underline{\underline{\lambda}}_5 = \begin{pmatrix} 0 & 0 & -\mathrm{i} \\ 0 & 0 & 0 \\ \mathrm{i} & 0 & 0 \end{pmatrix}, \tag{4.1}$$
$$\underline{\underline{\lambda}}_6 = \begin{pmatrix} 0 & 0 & 0 \\ 0 & 0 & 1 \\ 0 & 1 & 0 \end{pmatrix}, \quad \underline{\underline{\lambda}}_7 = \begin{pmatrix} 0 & 0 & 0 \\ 0 & 0 & -\mathrm{i} \\ 0 & \mathrm{i} & 0 \end{pmatrix}.$$

For the remaining diagonal λ-matrix we don't choose the form (2.6) because it does not meet the condition (2.27) for all pairs $\underline{\underline{\lambda}}_i$, $\underline{\underline{\lambda}}_k$. Referring to Gell–Mann, 1964, we use

$$\underline{\underline{\lambda}}_8 = \frac{1}{\sqrt{3}} \begin{pmatrix} 1 & 0 & 0 \\ 0 & 1 & 0 \\ 0 & 0 & -2 \end{pmatrix}. \tag{4.2}$$

Now the condition $Tr\left(\underline{\underline{\lambda}}_i\underline{\underline{\lambda}}_k\right) = 2\delta_{ik}$, (2.27), is satisfied completely. The matrix $\underline{\underline{\lambda}}_8$ is made up as a linear combination of basis elements of the type (2.6) like this:

$$\underline{\underline{\lambda}}_8 = \frac{1}{\sqrt{3}} \begin{pmatrix} 1 & 0 & 0 \\ 0 & -1 & 0 \\ 0 & 0 & 0 \end{pmatrix} + \frac{2}{\sqrt{3}} \begin{pmatrix} 0 & 0 & 0 \\ 0 & 1 & 0 \\ 0 & 0 & -1 \end{pmatrix}. \tag{4.3}$$

The structure constants C_{ikl}, which belong to the basis $\left\{\underline{\underline{e}}_i\right\}$, (see (2.25)), appear also in the following commutator relation:

$$\left[\underline{\underline{\lambda}}_i, \underline{\underline{\lambda}}_k\right] = \sum_l C_{ikl} 2\mathrm{i} \underline{\underline{\lambda}}_l \quad \text{(see (2.26))}. \tag{4.4}$$

They can be calculated by (2.29) or by elementary matrix multiplication and reduction. One obtains

Table 4.1.1. Table of the non-vanishing structure constants of $su(3)$.

i	k	l	C_{ikl}
1	2	3	1
1	4	7	1/2
1	5	6	−1/2
2	4	6	1/2
2	5	7	1/2
3	4	5	1/2
3	6	7	−1/2
4	5	8	$\sqrt{3}/2$
6	7	8	$\sqrt{3}/2$.

Due to (2.30) the structure constants are totally antisymmetric. Therefore, the remaining non-vanishing values can be obtained by permuting the indices listed above. Table 4.1.1 reveals that the following types of structure constants vanish: $C_{18x} = C_{28x} = C_{38x} = 0$ with $x = 1, 2, \ldots, 7$. That is,

$$\left[\underline{\underline{\lambda}}_1, \underline{\underline{\lambda}}_8\right] = \left[\underline{\underline{\lambda}}_2, \underline{\underline{\lambda}}_8\right] = \left[\underline{\underline{\lambda}}_3, \underline{\underline{\lambda}}_8\right] = 0. \tag{4.5}$$

On the other hand, Table 4.1.1 yields $\left[\underline{\underline{\lambda}}_1, \underline{\underline{\lambda}}_2\right] \neq 0$, $\left[\underline{\underline{\lambda}}_1, \underline{\underline{\lambda}}_3\right] \neq 0$, $\left[\underline{\underline{\lambda}}_2, \underline{\underline{\lambda}}_3\right] \neq 0$. That is, the matrices $\underline{\underline{\lambda}}_1$, $\underline{\underline{\lambda}}_2$, $\underline{\underline{\lambda}}_3$, and $\underline{\underline{\lambda}}_8$ do not mutually commute. Investigations of this kind show that in $su(3)$ there are not more than two $\underline{\underline{\lambda}}$'s which mutually commute. Therefore, one says that the $su(3)$-algebra has *rank* 2.

For most pairs (i, k) in equation (4.4) exists only one l-value. The pairs (4,5) and (6,7) result in two summands.

4.2 Subalgebras of the $su(3)$-algebra

In (2.22) we have introduced the operators $\boldsymbol{\lambda}_i$ by the relation $\boldsymbol{e}_i = -\frac{1}{2}\mathrm{i}\boldsymbol{\lambda}_i$. Here we deal with the set of those $\boldsymbol{\lambda}_i$-operators that constitute also a Lie algebra, which is isomorphic to the algebra formed by the set $\left\{\underline{\underline{\lambda}}_i\right\}$ of (4.1) and (4.2).

In addition, using the $\boldsymbol{\lambda}_i$'s, we can make up **basis operators of the $su(2)$-algebras** like this:

$$\begin{array}{llll} \boldsymbol{e}_1^{(T)} = -\frac{1}{2}\mathrm{i}\boldsymbol{\lambda}_1, & \boldsymbol{e}_2^{(T)} = -\frac{1}{2}\mathrm{i}\boldsymbol{\lambda}_2, & \boldsymbol{e}_3^{(T)} = -\frac{1}{2}\mathrm{i}\boldsymbol{\lambda}_3 & \text{and} \\ \boldsymbol{e}_1^{(U)} = -\frac{1}{2}\mathrm{i}\boldsymbol{\lambda}_6, & \boldsymbol{e}_2^{(U)} = -\frac{1}{2}\mathrm{i}\boldsymbol{\lambda}_7, & \boldsymbol{e}_3^{(U)} = -\frac{1}{4}\mathrm{i}\left(-\boldsymbol{\lambda}_3 + \sqrt{3}\boldsymbol{\lambda}_8\right) & \text{and} \\ \boldsymbol{e}_1^{(V)} = -\frac{1}{2}\mathrm{i}\boldsymbol{\lambda}_4, & \boldsymbol{e}_2^{(V)} = -\frac{1}{2}\mathrm{i}\boldsymbol{\lambda}_5, & \boldsymbol{e}_3^{(V)} = -\frac{1}{4}\mathrm{i}\left(\boldsymbol{\lambda}_3 + \sqrt{3}\boldsymbol{\lambda}_8\right). & \end{array} \tag{4.6}$$

Using (4.4) and Table 4.1.1, one proves that for each triple of $\boldsymbol{e}$-operators the relation (3.25)

$$\left[\boldsymbol{e}_1^{(X)}, \boldsymbol{e}_2^{(X)}\right] = \boldsymbol{e}_3^{(X)}, \quad \text{cyclic in } 1, 2, 3, \quad \text{for } X = T,\ U \text{ or } V, \tag{4.7}$$

is satisfied in accordance with (3.2). That is, each triple constitutes a $su(2)$-algebra. It is a subalgebra of $su(3)$.

These **subalgebras** are **not invariant** (see Subsection 1.5). We check this for the first one like this:

$$\left[\boldsymbol{e}_1^{(T)}, \boldsymbol{e}_4\right] = -\frac{1}{4}\left[\boldsymbol{\lambda}_1, \boldsymbol{\lambda}_4\right] = -\frac{1}{4} \cdot \frac{1}{2} \cdot 2\,\mathrm{i}\boldsymbol{\lambda}_7 = \frac{1}{2}\boldsymbol{e}_7. \tag{4.8}$$

This element is not a member of the first subalgebra. Therefore this is not invariant. There are no invariant subalgebras in $su(3)$ and therefore this algebra is *simple* like every $su(N)$-algebra — as mentioned in Section 2.2.

In analogy with (3.41), we introduce the following **step operators and "third" operators**

$$\boldsymbol{T}_{\pm} = \frac{1}{2}(\boldsymbol{\lambda}_1 \pm \mathrm{i}\boldsymbol{\lambda}_2) \quad \text{and} \quad \boldsymbol{T}_3 = \frac{1}{2}\boldsymbol{\lambda}_3. \tag{4.9}$$

We obtain from (4.4) and Table 4.1.1

$$[\boldsymbol{T}_3, \boldsymbol{T}_{\pm}] = \pm\boldsymbol{T}_{\pm}, \quad [\boldsymbol{T}_+, \boldsymbol{T}_-] = 2\boldsymbol{T}_3 \tag{4.10}$$

in accordance with (3.43). Defining the operators

$$\begin{aligned} \boldsymbol{U}_{\pm} &= \tfrac{1}{2}(\boldsymbol{\lambda}_6 \pm \mathrm{i}\boldsymbol{\lambda}_7), \quad \boldsymbol{U}_3 = \tfrac{1}{4}\left(-\boldsymbol{\lambda}_3 + \sqrt{3}\boldsymbol{\lambda}_8\right), \\ \boldsymbol{V}_{\pm} &= \tfrac{1}{2}(\boldsymbol{\lambda}_4 \pm \mathrm{i}\boldsymbol{\lambda}_5), \quad \boldsymbol{V}_3 = \tfrac{1}{4}\left(\boldsymbol{\lambda}_3 + \sqrt{3}\boldsymbol{\lambda}_8\right), \end{aligned} \tag{4.11}$$

the following commutation relations result, in analogy to (4.10):

$$\begin{aligned} [\boldsymbol{U}_3, \boldsymbol{U}_{\pm}] &= \pm\boldsymbol{U}_{\pm}, \quad [\boldsymbol{U}_+, \boldsymbol{U}_-] = 2\boldsymbol{U}_3, \\ [\boldsymbol{V}_3, \boldsymbol{V}_{\pm}] &= \pm\boldsymbol{V}_{\pm}, \quad [\boldsymbol{V}_+, \boldsymbol{V}_-] = 2\boldsymbol{V}_3. \end{aligned} \tag{4.12}$$

Among some of the nine operators in (4.9) and (4.11), the following linear dependency exists:

$$\boldsymbol{U}_3 = -\boldsymbol{T}_3 + \boldsymbol{V}_3, \tag{4.13}$$

which we expect because we started with eight $\boldsymbol{\lambda}$'s. The correspondence between the commutation relations (4.10) and (3.43) shows that $su(2)$-multiplets are affiliated to the algebra $\{\boldsymbol{T}_+, \boldsymbol{T}_-, \boldsymbol{T}_3\}$ (see the end of Section 3.3). This is also true for the $\boldsymbol{U}$- and the $\boldsymbol{V}$-algebra. In Table 4.2.1 the commutator relations of the $\boldsymbol{T}-$, $\boldsymbol{U}-$ and $\boldsymbol{V}-$operators are put together.

Table 4.2.1. Commutator relations in $su(3)$. The commutator of an operator in the top line with an operator in the column at the right hand side results in the operator given in the corresponding field.

$\boldsymbol{T}_-$	$\boldsymbol{T}_3$	$\boldsymbol{U}_+$	$\boldsymbol{U}_-$	$\boldsymbol{U}_3$	$\boldsymbol{V}_+$	$\boldsymbol{V}_-$	$\boldsymbol{V}_3$	
$-2\boldsymbol{T}_3$	$\boldsymbol{T}_+$	$-\boldsymbol{V}_+$	0	$-\frac{1}{2}\boldsymbol{T}_+$	0	$\boldsymbol{U}_-$	$\frac{1}{2}\boldsymbol{T}_+$	$\boldsymbol{T}_+$
	$-\boldsymbol{T}_-$	0	$\boldsymbol{V}_-$	$\frac{1}{2}\boldsymbol{T}_-$	$-\boldsymbol{U}_+$	0	$-\frac{1}{2}\boldsymbol{T}_-$	$\boldsymbol{T}_-$
		$\frac{1}{2}\boldsymbol{U}_+$	$-\frac{1}{2}\boldsymbol{U}_-$	0	$-\frac{1}{2}\boldsymbol{V}_+$	$\frac{1}{2}\boldsymbol{V}_-$	0	$\boldsymbol{T}_3$
			$-2\boldsymbol{U}_3$	$\boldsymbol{U}_+$	0	$-\boldsymbol{T}_-$	$\frac{1}{2}\boldsymbol{U}_+$	$\boldsymbol{U}_+$
				$-\boldsymbol{U}_-$	$\boldsymbol{T}_+$	0	$-\frac{1}{2}\boldsymbol{U}_-$	$\boldsymbol{U}_-$
					$-\frac{1}{2}\boldsymbol{V}_+$	$\frac{1}{2}\boldsymbol{V}_-$	0	$\boldsymbol{U}_3$
						$-2\boldsymbol{V}_3$	$\boldsymbol{V}_+$	$\boldsymbol{V}_+$
							$-\boldsymbol{V}_-$	$\boldsymbol{V}_-$

In the lower triplets the relations (4.10) and (4.12) are visible. The fact that $\boldsymbol{T}_3$, $\boldsymbol{U}_3$ and $\boldsymbol{V}_3$ mutually commute does not influence the statement that $su(3)$ has rank 2 (see end of Section 4.1) because these operators are linearly dependent (see (4.13)).

4.3 Step operators and states in $su(3)$

Corresponding to (3.39), we can achieve that every state (function) of $su(3)$ is an eigenstate of $\boldsymbol{T}_3$. On the other hand, it might be an eigenstate of $\boldsymbol{U}_3$ or $\boldsymbol{V}_3$. However, these three operators mutually commute (see Section 4.2) and therefore, every state is simultaneously eigenstate of $\boldsymbol{T}_3$, $\boldsymbol{U}_3$ and $\boldsymbol{V}_3$ in a certain analogy with the relation (3.40). Consequently, we can mark a $su(3)$-state by

$$|T_3\, U_3\, V_3\rangle\,, \tag{4.14}$$

where T_3, U_3 and V_3 are eigenvalues of $\boldsymbol{T}_3$, $\boldsymbol{U}_3$ and $\boldsymbol{V}_3$. The inner product

$$\langle T_3\, U_3\, V_3|\, \boldsymbol{T}_3\, |T_3'\, U_3'\, V_3'\rangle = T_3\, \delta_{T_3\, T_3'}\, \delta_{U_3\, U_3'}\, \delta_{V_3\, V_3'} \tag{4.15}$$

with analogue expressions for $\boldsymbol{U}_3$ and $\boldsymbol{V}_3$ reveals that $\boldsymbol{T}_3$, $\boldsymbol{U}_3$ and $\boldsymbol{V}_3$ can be diagonalised simultaneously — a formulation which is often used in this situation. In short, we can write

$$\begin{aligned} \boldsymbol{T}_3\, |T_3\, U_3\, V_3\rangle &= T_3\, |T_3\, U_3\, V_3\rangle\,, \\ \boldsymbol{U}_3\, |T_3\, U_3\, V_3\rangle &= U_3\, |T_3\, U_3\, V_3\rangle\,, \\ \boldsymbol{V}_3\, |T_3\, U_3\, V_3\rangle &= V_3\, |T_3\, U_3\, V_3\rangle\,. \end{aligned} \tag{4.16}$$

We apply the equation $[\boldsymbol{T}_3, \boldsymbol{T}_\pm] = \pm \boldsymbol{T}_\pm$, (4.10), to (4.14), which results in

$$\begin{aligned} \boldsymbol{T}_3\, (\boldsymbol{T}_\pm\, |T_3\, U_3\, V_3\rangle) &= \boldsymbol{T}_\pm \boldsymbol{T}_3\, |T_3\, U_3\, V_3\rangle \pm \boldsymbol{T}_\pm\, |T_3\, U_3\, V_3\rangle \\ &= (T_3 \pm 1)\, (\boldsymbol{T}_\pm\, |T_3\, U_3\, V_3\rangle)\,, \end{aligned} \tag{4.17}$$

in a certain analogy with (3.47). Therefore $\boldsymbol{T}_\pm\, |T_3\, U_3\, V_3\rangle$ is an eigenstate of $\boldsymbol{T}_3$ with the eigenvalue $T_3 \pm 1$. That is, we can expand

$$\boldsymbol{T}_\pm\, |T_3\, U_3\, V_3\rangle = \sum_{U_3'\, V_3'} N\, (T_3\, U_3\, V_3\, U_3'\, V_3')\;\, |T_3 \pm 1,\, U_3'\, V_3'\rangle\,. \tag{4.18}$$

Due to

$$[\boldsymbol{U}_3, \boldsymbol{T}_\pm] = \mp \frac{1}{2} \boldsymbol{T}_\pm \;\text{ (see Table 4.2.1)} \tag{4.19}$$

we obtain the relation

$$\begin{aligned}
\boldsymbol{U}_3\left(\boldsymbol{T}_\pm\left|T_3\,U_3\,V_3\right\rangle\right) &= \boldsymbol{T}_\pm\boldsymbol{U}_3\left|T_3\,U_3\,V_3\right\rangle \mp \frac{1}{2}\boldsymbol{T}_\pm\left|T_3\,U_3\,V_3\right\rangle \\
&= \left(U_3 \mp \frac{1}{2}\right)\left(\boldsymbol{T}_\pm\left|T_3\,U_3\,V_3\right\rangle\right), \qquad (4.20)\\
\text{i.e.,}\qquad \boldsymbol{T}_\pm\left|T_3\,U_3\,V_3\right\rangle &= \sum_{T_3''\,V_3''} N\left(T_3\,U_3\,V_3\,T_3''\,V_3''\right)\left|T_3'', U_3 \mp \frac{1}{2}, V_3''\right\rangle .
\end{aligned}$$

Analogously, from

$$[\boldsymbol{V}_3, \boldsymbol{T}_\pm] = \pm\frac{1}{2}\boldsymbol{T}_\pm \tag{4.21}$$

we get

$$\boldsymbol{T}_\pm\left|T_3\,U_3\,V_3\right\rangle = \sum_{T_3'''\,U_3'''} N\left(T_3\,U_3\,V_3\,T_3'''\,U_3'''\right)\left|T_3''', U_3''', V_3 \pm \frac{1}{2}\right\rangle . \tag{4.22}$$

That is to say, acting on $|T_3\,U_3\,V_3\rangle$ the operator $\boldsymbol{T}_\pm$ generates a state with quantum numbers $T_3 \pm 1$, $U_3 \mp \frac{1}{2}$ and $V_3 \pm \frac{1}{2}$. We write

$$\boldsymbol{T}_\pm\left|T_3\,U_3\,V_3\right\rangle \rightarrow \left|T_3 \pm 1,\ U_3 \mp \frac{1}{2},\ V_3 \pm \frac{1}{2}\right\rangle . \tag{4.23}$$

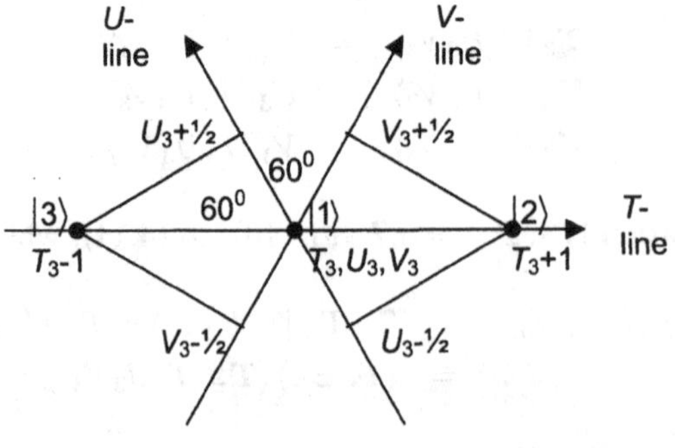

Figure 4.3.1. T-, U- and V-lines.
Representation of the U- and V- components of states lying on the T-axis.

This behaviour is realised graphically in Figure 4.3.1 by drawing an axis for the T-subalgebra analogously to Figure 3.7.1, an axis for the U- and an other for the V-subalgebra. Their directions differ by 60^0. Assuming the coordinates T_3, U_3 and V3 for state $|1\rangle$ the operator T_+ creates the state $T_+\,|1\rangle = |2\rangle$ (definition) on the T-line with the coordinate T_3+1. Its U-coordinate is found by dropping a perpendicular on the U-line resulting in U_3 - 1/2 and on the V-line yielding V_3 + 1/2 in accordance with (4.23). The action of T_- can be read in the same way from Figure 4.3.1. Analogously to (4.23) the following relations are derived:

$$\begin{aligned} \boldsymbol{U}_\pm \left|T_3 U_3 V_3\right\rangle &\rightarrow \left|T_3 \mp \tfrac{1}{2}, U_3 \pm 1, V_3 \pm \tfrac{1}{2}\right\rangle, \\ \boldsymbol{V}_\pm \left|T_3 U_3 V_3\right\rangle &\rightarrow \left|T_3 \pm \tfrac{1}{2}, U_3 \pm \tfrac{1}{2}, V_3 \pm 1\right\rangle. \end{aligned} \tag{4.24}$$

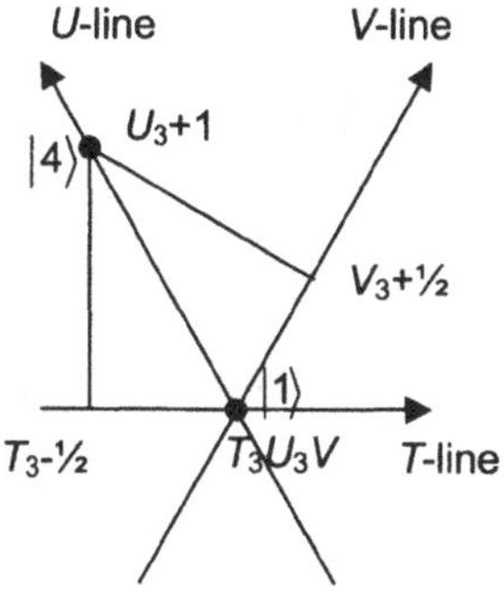

Figure 4.3.2. Representation of the T- and V-component of a state lying on the U-axis.

In Figure 4.3.2 the first relation of (4.24) is reproduced correctly, because the state $\boldsymbol{U}_+ \left|1\right\rangle = \left|4\right\rangle$ (definition) shows $U_3 + 1$ and the components $V_3 + 1/2$ and $T_3 - 1/2$. The second expression in (4.24) becomes apparent in an analogue graphical scheme.

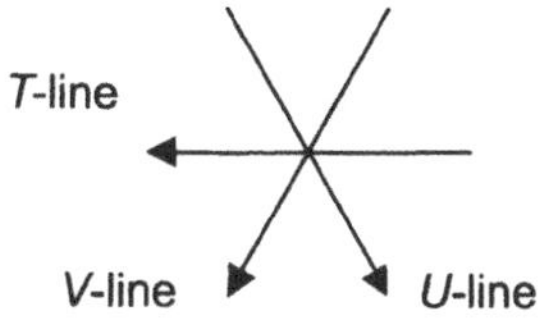

Figure 4.3.3. "Anti"-coordinate system of $su(3)$

Of course the "anti"-coordinate system of Figure 4.3.3 with reversed T-, U- and V-lines represents the relations (4.23) and (4.24) in the same way. In Sections 4.8 and 4.10 we will refer to it.

4.4 Multiplets of $su(3)$

Repeated applications of the operators $\boldsymbol{T}_\pm, \boldsymbol{U}_\pm$ and $\boldsymbol{V}_\pm$ generate a lattice of states, which are shown in Figure 4.4.1 as points in a triangular formation.

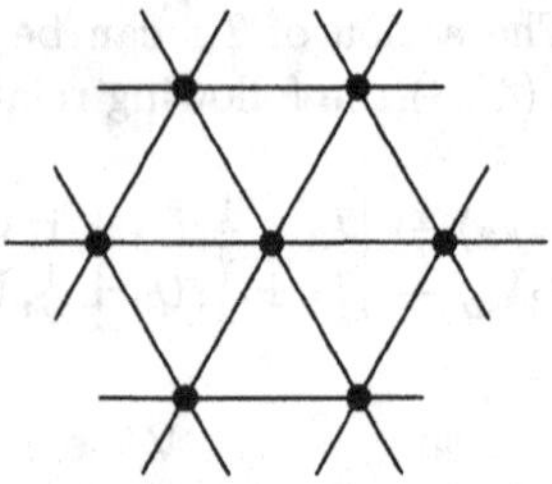

Figure 4.4.1. Lattice of $su(3)$-states.

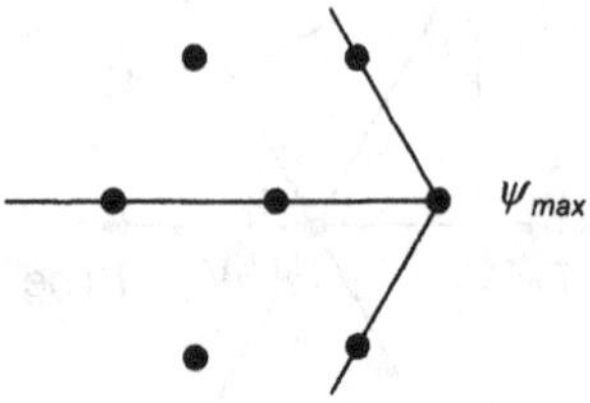

Figure 4.4.2. Maximal weight of the $su(3)$-multiplet.

We demand that $su(3)$-multiplets are finite. We start from the state with the highest T_3-value, which we name $T_{3,\max}$ or *maximal weight.* Its state is marked with $\psi_{\max}$ (see Figure 4.4.2) and it must meet the relation $\boldsymbol{T}_+\psi_{\max} = 0$. Due to the hexagonal (or triangular) structure of the point lattice, the statements

$$\boldsymbol{U}_-\psi_{\max} = \boldsymbol{V}_+\psi_{\max} = 0 \tag{4.25}$$

are true as well.

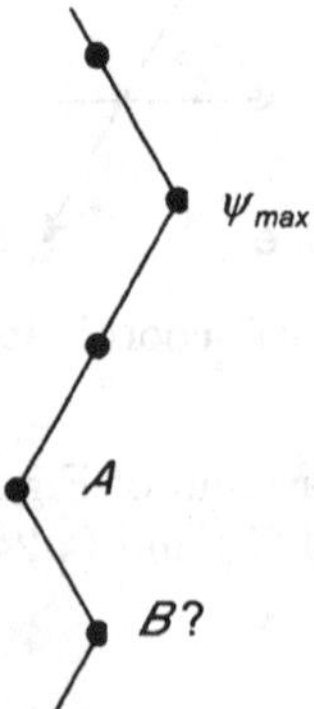

Figure 4.4.3. Hypothesis of a concave boundary.

Now we move on $n \geq 1$ steps from the state of maximal weight in the V-line and reach point A, where the state $\psi_A = (\boldsymbol{V}_-)^n\,\psi_{\max}$ exists (in Figure 4.4.3, $n = 2$

is chosen). If we would assume that the boundary of the multiplet is concave in A and if there would be one step up to the point B, the following relation would hold:

$$\psi_B = \boldsymbol{U}_- \psi_A = \boldsymbol{U}_- \left(\boldsymbol{V}_-\right)^n \psi_{\mathrm{max}} = \left(\boldsymbol{V}_-\right)^n \boldsymbol{U}_- \psi_{\mathrm{max}} = 0, \tag{4.26}$$

where we have made use of $[\boldsymbol{U}_-, \boldsymbol{V}_-] = 0$, (Table 4.2.1), and of (4.25). Therefore, the state B and the concave corner in A do not exist. In the same way, one shows that

The su(3)-multiplet is convex everywhere in the T-U-V-plane.

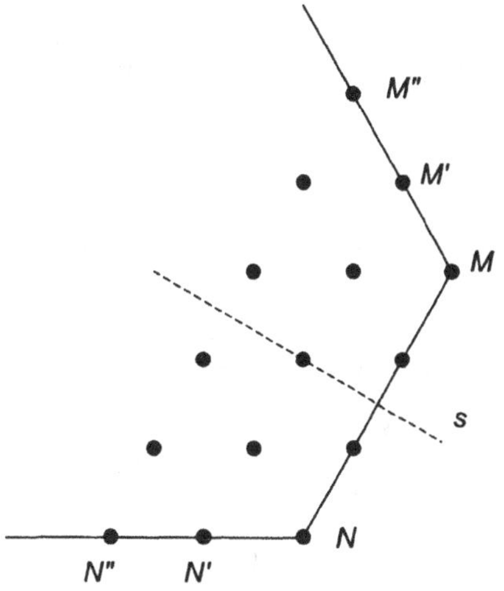

Figure 4.4.4. Part of a $su(3)$-multiplet.

We refer to Figure 4.4.4, where a part of a $su(3)$-multiplet is outlined. The point chain between M and N represents a $su(2)$-multiplet in V-direction (according to Figure 3.7.1), which has a natural symmetry. Correspondingly, the neighbouring points of M must have a symmetrical counterpart in the neighbourhood of N. That is, the $su(2)$-multiplet $N' - M'$ is also symmetric with respect to the perpendicular s. This is true for the other V-multiplets (the $N'' - M''$-multiplet, for instance). Therefore, the U-line MM'' has the same length as the T-line NN''.

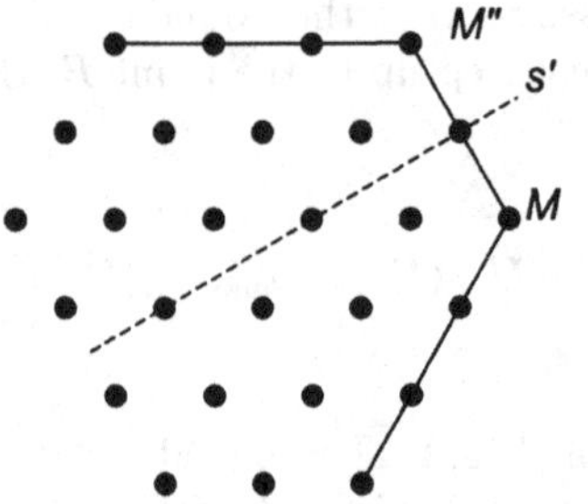

Figure 4.4.5. Second symmetry axis of the $su(3)$-multiplet.

Now we start from the symmetry of the U-multiplet MM'' (see Figure 4.4.5). The remaining parallel submultiplets must also be symmetric with respect to the perpendicular s'. In this way the area of the $su(3)$-multiplet is closed forming a partially regular hexagon with three symmetry axes including angles of 60^0 (see Figure 4.4.6). $p+1$ is the total number of state points on the V-line going through the point with maximal weight, and $q+1$ is the analogue number for the corresponding U-line. The integers p and q characterise the $su(3)$-multiplet completely. It is marked by $M(p,q)$. Thus, in Figure 4.4.6 we have the $su(3)$-multiplet $M(3,2)$.

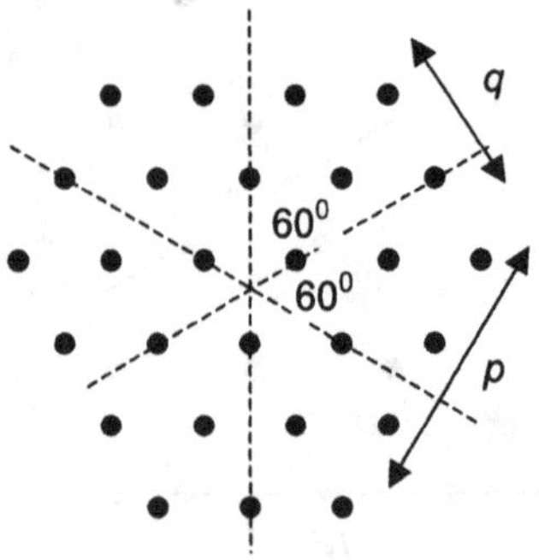

Figure 4.4.6. Ordering numbers of the $su(3)$-multiplets.

4.5 Individual states of the $su(3)$-multiplet and their multiplicities

First we try to characterise the state on point M carrying the maximal weight (see Figure 4.4.5). The submultiplet on the V-line going through M has $p+1$ states. Because it is a $su(2)$-multiplet we can write

$$p + 1 = 2j_V + 1. \tag{4.27}$$

The point M is a furthest member of this multiplet. Its magnetic quantum number is therefore $m_{V,\max} = j_V$ and the state can be described by

$$|M\rangle = |j_V, m_{V,\max}\rangle = \left|\frac{p}{2}, \frac{p}{2}\right\rangle_V . \tag{4.28}$$

On the other hand, the state $|M\rangle$ is part of a U-multiplet. Accordingly,

$$|M\rangle = \left|\frac{q}{2}, -\frac{q}{2}\right\rangle_U \tag{4.29}$$

holds. Because M is in the largest T-multiplet (see Figure 4.5.1) and shows $j_T = \frac{p}{2} + \frac{q}{2} = m_{T,\max}$, we can write

$$|M\rangle = \left|\frac{p+q}{2}, \frac{p+q}{2}\right\rangle_T . \tag{4.30}$$

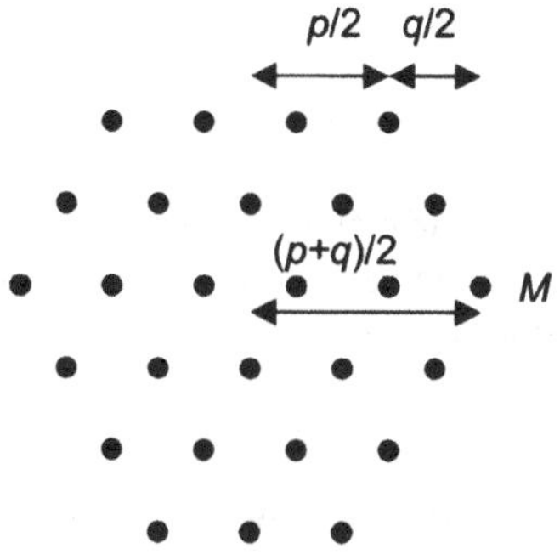

Figure 4.5.1. The largest T-multiplet.

In (4.28) up to (4.30) the relative phases +1 have been chosen. The state $|M\rangle$ is called *unique*, because it can be described by one expression of the T-, U- or V-type.

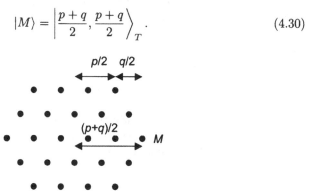

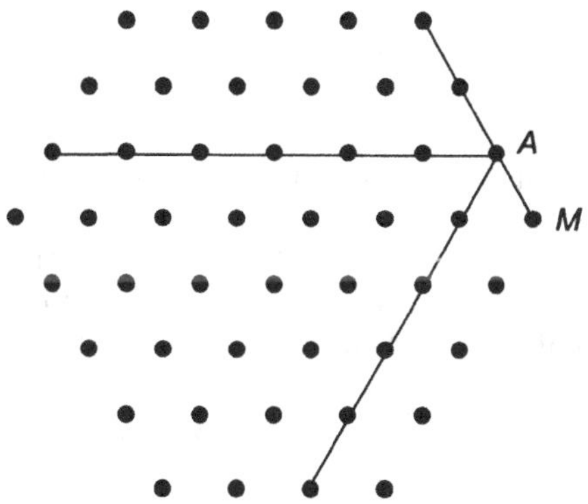

Figure 4.5.2. Submultiplets of a boundary state.

We claim that the neighbouring state $|A\rangle$ (see Figure 4.5.2) is also unique. It lies at the end of a V-multiplet with $p+2$ states and is described by

$$|A\rangle = \left|\frac{p+1}{2}, \frac{p+1}{2}\right\rangle_V, \tag{4.31}$$

and from the corresponding T-multiplet we learn

$$|A\rangle = \left|\frac{p+q-1}{2}, \frac{p+q-1}{2}\right\rangle_T. \tag{4.32}$$

We define: *a state is unique if it is represented at least in one subalgebra (T-, U- or V) by a single expression.* In the U-subalgebra for $|A\rangle$ the states

$$\left|\frac{q}{2}, -\frac{q}{2}+1\right\rangle_U \quad \text{and} \tag{4.33}$$

$$\left|\frac{q}{2}-1, -\frac{q}{2}+1\right\rangle_U \tag{4.34}$$

are possible. But we know already that $|A\rangle$ is unique, for which reason we drop the last expression (4.34). In the same way one proves that all the states on the boundary of a $su(3)$-multiplet are unique.

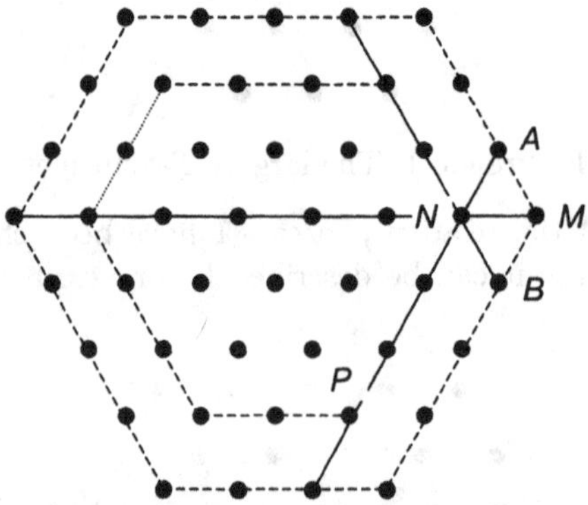

Figure 4.5.3. A state on the inner shell.

Now we deal with the state $|N\rangle$ in Figure 4.5.3 and claim that it is **doubly occupied**. Double occupation denotes that of the three descriptions of states (in the T-multiplet, in the U-multiplet and in the V-multiplet) one is composed out of *a pair of orthonormalised su(2)-states.* The other two descriptions may comprise two or more $su(2)$-states each. Similarly to (4.33/34) the U-expressions

$$\left|\frac{q+1}{2}, -\frac{q-1}{2}\right\rangle_U, \left|\frac{q-1}{2}, -\frac{q-1}{2}\right\rangle_U \tag{4.35}$$

can be assigned to $|N\rangle$, or, in other words, the state $|N\rangle$ can be written as a linear combination of these orthogonal expressions. Analogously,

$$\begin{array}{l} \left|\frac{p+q}{2},\frac{p+q-2}{2}\right\rangle_T, \left|\frac{p+q-2}{2},\frac{p+q-2}{2}\right\rangle_T \text{ and} \\ \left|\frac{p+1}{2},\frac{p-1}{2}\right\rangle_V, \left|\frac{p-1}{2},\frac{p-1}{2}\right\rangle_V \end{array} \tag{4.36}$$

belong to $|N\rangle$. Obviously, $|N\rangle$ is doubly occupied. The pairs of expressions in (4.35) and (4.36) are equivalent in such a way that the members of one pair, say the T-pair, can be expressed linearly by the members of an other pair, say the U-pair. We carry it out this way:

$$\begin{array}{l} \left|\frac{p+q}{2},\frac{p+q-2}{2}\right\rangle_T = a_1\left|\frac{q+1}{2},-\frac{q-1}{2}\right\rangle_U + a_2\left|\frac{q-1}{2},-\frac{q-1}{2}\right\rangle_U \\ \left|\frac{p+q-2}{2},\frac{p+q-2}{2}\right\rangle_T = a_2\left|\frac{q+1}{2},-\frac{q-1}{2}\right\rangle_U - a_1\left|\frac{q-1}{2},-\frac{q-1}{2}\right\rangle_U . \end{array} \tag{4.37}$$

In the second equation the coefficients a_2 and $-a_1$ have been chosen in order to make both lines of (4.37) orthogonal to each other, which we see like this:

$$0 = \left\langle \frac{p+q-2}{2},\frac{p+q-2}{2} \,\middle|\, \frac{p+q}{2},\frac{p+q-2}{2} \right\rangle_T = a_2a_1 - a_1a_2 = 0. \tag{4.38}$$

Analogously to (4.37), the following relations hold:

$$\begin{array}{l} \left|\frac{p+q}{2},\frac{p+q-2}{2}\right\rangle_T = b_1\left|\frac{p+1}{2},\frac{p-1}{2}\right\rangle_V + b_2\left|\frac{p-1}{2},\frac{p-1}{2}\right\rangle_V \\ \left|\frac{p+q-2}{2},\frac{p+q-2}{2}\right\rangle_T = b_2\left|\frac{p+1}{2},\frac{p-1}{2}\right\rangle_V - b_1\left|\frac{p-1}{2},\frac{p-1}{2}\right\rangle_V . \end{array} \tag{4.39}$$

In order to determine the coefficients a_i and b_i in (4.37) and (4.39) we make use of the relation $\boldsymbol{U}_+\boldsymbol{V}_- - \boldsymbol{V}_-\boldsymbol{U}_+ - \boldsymbol{T}_- = 0$ (see Table 4.2.1), apply this null-operator to the state $|M\rangle$ and insert the expressions (4.28), (4.29) or (4.30) this way:

$$\begin{aligned} 0 &= (\boldsymbol{U}_+\boldsymbol{V}_- - \boldsymbol{V}_-\boldsymbol{U}_+ - \boldsymbol{T}_-)\,|M\rangle \\ &= \boldsymbol{U}_+\boldsymbol{V}_-\left|\frac{p}{2}\frac{p}{2}\right\rangle_V - \boldsymbol{V}_-\boldsymbol{U}_+\left|\frac{q}{2},-\frac{q}{2}\right\rangle_U - \boldsymbol{T}_-\left|\frac{p+q}{2},\frac{p+q}{2}\right\rangle_T . \end{aligned} \tag{4.40}$$

The $su(2)$ step operators act according to (3.65) and (3.66) (remember, the commutator relations for the operators $\boldsymbol{J}_+,\boldsymbol{J}_-,\boldsymbol{J}_3$, (3.43), are identical to these relations for $\boldsymbol{T}_+,\boldsymbol{T}_-,\boldsymbol{T}_3$, (4.10), and for $\boldsymbol{V}_+,\boldsymbol{V}_-,\boldsymbol{V}_3$ and $\boldsymbol{U}_+,\boldsymbol{U}_-,\boldsymbol{U}_3$, (4.12)). We obtain

$$\begin{aligned}
0 &= U_+\sqrt{\frac{p}{2}\cdot\frac{p+2}{2}-\frac{p}{2}\cdot\frac{p-2}{2}}\,|B\rangle - V_-\sqrt{\frac{q}{2}\cdot\frac{q+2}{2}-\frac{q}{2}\cdot\frac{q-2}{2}}\,|A\rangle \\
&\quad -\sqrt{\frac{p+q}{2}\cdot\frac{p+q+2}{2}-\frac{p+q}{2}\cdot\frac{p+q-2}{2}}\left|\frac{p+q}{2},\frac{p+q-2}{2}\right\rangle_T \\
&= \sqrt{p}\sqrt{\frac{q+1}{2}\cdot\frac{q+3}{2}-\frac{q+1}{2}\cdot\frac{q-1}{2}}\left|\frac{q+1}{2},-\frac{q-1}{2}\right\rangle_U \\
&\quad -\sqrt{q}\sqrt{\frac{p+1}{2}\cdot\frac{p+3}{2}-\frac{p+1}{2}\cdot\frac{p-1}{2}}\left|\frac{p+1}{2},\frac{p-1}{2}\right\rangle_V \\
&\quad -\sqrt{p+q}\left|\frac{p+q}{2},\frac{p+q-2}{2}\right\rangle_T \\
&= \sqrt{p\,(q+1)}\left|\frac{q+1}{2},-\frac{q-1}{2}\right\rangle_U - \sqrt{q\,(p+1)}\left|\frac{p+1}{2},\frac{p-1}{2}\right\rangle_V \qquad (4.41)\\
&\quad -\sqrt{p+q}\left|\frac{p+q}{2},\frac{p+q-2}{2}\right\rangle_T,
\end{aligned}$$

where we have used the relations (4.31), (4.34) and $|B\rangle = \left|\frac{p+q+1}{2},\frac{p+q-1}{2}\right\rangle_T = \left|\frac{q+1}{2},-\frac{q+1}{2}\right\rangle_U = \left|\frac{p}{2},\frac{p-2}{2}\right\rangle_V$ (Note: here the states $|A\rangle, |B\rangle$ on the boundary are not normalised in the same way as in Sections 4.3 and 4.4). We now form the inner product of (4.41) with the bra state ${}_T\left\langle\frac{p+q}{2},\frac{p+q-2}{2}\right|$, i.e.,

$$\begin{aligned}
&\sqrt{p\,(q+1)}\;{}_T\left\langle\tfrac{p+q}{2},\tfrac{p+q-2}{2}\;\middle|\;\tfrac{q+1}{2},-\tfrac{q-1}{2}\right\rangle_U \\
&-\sqrt{q\,(p+1)}\;{}_T\left\langle\tfrac{p+q}{2},\tfrac{p+q-2}{2}\;\middle|\;\tfrac{p+1}{2},\tfrac{p-1}{2}\right\rangle_V \\
&-\sqrt{p+q}\;{}_T\left\langle\tfrac{p+q}{2},\tfrac{p+q-2}{2}\;\middle|\;\tfrac{p+q}{2},\tfrac{p+q-2}{2}\right\rangle_T = 0.
\end{aligned} \qquad (4.42)$$

We insert (4.37) and (4.39) and make use of the orthonormality of the $su(2)$-states:

$$\begin{aligned}
&a_1\sqrt{p\,(q+1)}\;{}_U\left\langle\tfrac{q+1}{2},-\tfrac{q-1}{2}\;\middle|\;\tfrac{q+1}{2},-\tfrac{q-1}{2}\right\rangle_U \\
&-b_1\sqrt{q\,(p+1)}\;{}_V\left\langle\tfrac{p+1}{2},\tfrac{p-1}{2}\;\middle|\;\tfrac{p+1}{2},\tfrac{p-1}{2}\right\rangle_V \\
&-\sqrt{p+q}\;{}_T\left\langle\tfrac{p+q}{2},\tfrac{p+q-2}{2}\;\middle|\;\tfrac{p+q}{2},\tfrac{p+q-2}{2}\right\rangle_T = 0.
\end{aligned} \qquad (4.43)$$

That is,

$$a_1\sqrt{p\,(q+1)} - b_1\sqrt{q\,(p+1)} - \sqrt{p+q} = 0. \qquad (4.44)$$

Analogously, we make up the inner product of (4.41) with ${}_T\left\langle\frac{p+q-2}{2},\frac{p+q-2}{2}\right|$ and get

$$a_2\sqrt{p\,(q+1)} - b_2\sqrt{q\,(p+1)} = 0. \qquad (4.45)$$

We bring the terms with b_i on the left-hand side, square and get from (4.44) and (4.45)

$$\begin{aligned}
b_1^2 q\,(p+1) &= a_1^2 p\,(q+1) - 2a_1\sqrt{p\,(q+1)\,(p+q)} + p + q \\
b_2^2 q\,(p+1) &= a_2^2 p\,(q+1)\,.
\end{aligned} \qquad (4.46)$$

Making use of $a_1^2 + a_2^2 = b_1^2 + b_2^2 = 1$, we obtain

$$a_1 = \sqrt{\frac{p}{(q+1)(p+q)}}, \quad a_2 = +\sqrt{1 - a_1^2} = \sqrt{\frac{q(p+q)+q}{(q+1)(p+q)}}. \tag{4.47}$$

Analogously, the equations (4.44) and (4.45) yield

$$b_1 = -\sqrt{\frac{q}{(q+1)(p+q)}}, \quad b_2 = \sqrt{\frac{q(p+q)+p}{(q+1)(p+q)}}. \tag{4.48}$$

For a_2 and b_2 positive square roots are chosen.

The states on the V-line between N and P in Figure 4.5.3 contain more than two V-expressions. But since there are only two U- and T-expressions these states are said to be doubly occupied, in analogy to the uniqueness of boundary states. Thus, *the states of the first inner shell* (next to the boundary shell) *are doubly occupied.* The following shell is triply occupied — provided that the preceding is hexagonal — and so on.

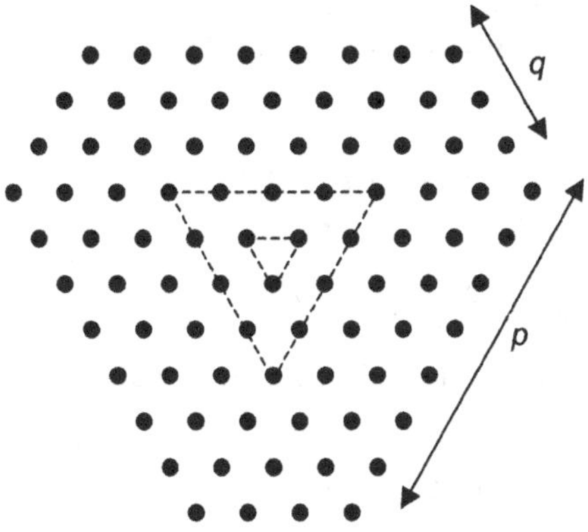

Figure 4.5.4. Hexagonal and triangular inner shells.

According to Figure 4.5.4 from the boundary one reaches a **triangular shell** after q steps, $(q < p)$. Each state on a face of the triangle has multiplicity $q+1$. We will show that from now on the multiplicity is not raised passing inward, i.e., all triangular shells have the same multiplicity. We deal with this situation starting with a $su(3)$-multiplet with a triangular boundary $(q = 0)$. For the states $|M\rangle$, $|A\rangle$ and $|\mathrm{B}\rangle$ in Figure 4.5.5, we write

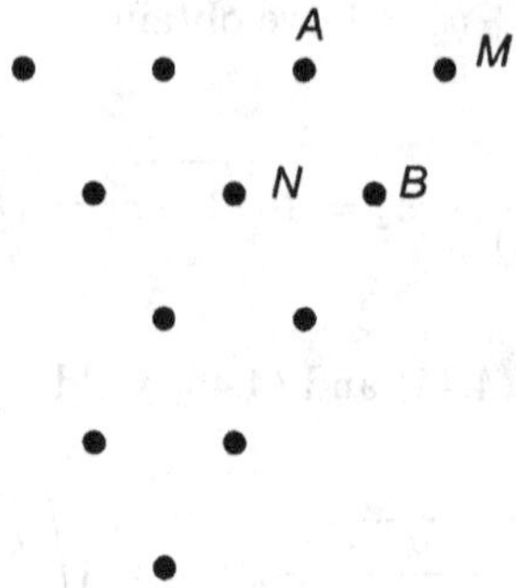

Figure 4.5.5. Part of a triangular multiplet.

$$\begin{aligned}
|M\rangle &= \left|\frac{p}{2},\frac{p}{2}\right\rangle_T = |0,0\rangle_U = \left|\frac{p}{2},\frac{p}{2}\right\rangle_V, \\
|A\rangle &= \left|\frac{p}{2},\frac{p-2}{2}\right\rangle_T = \left|\frac{p-1}{2},\frac{p-1}{2}\right\rangle_V, \\
|B\rangle &= \left|\frac{p-1}{2},\frac{p-1}{2}\right\rangle_T = \left|\frac{p}{2},\frac{p-2}{2}\right\rangle_V.
\end{aligned} \tag{4.49}$$

The states $|A\rangle$ and $|B\rangle$ lie on a U-line and are unique. Therefore, the additional V- and T-states are omitted. Due to Table 4.2.1, the operator relation $\boldsymbol{T}_-\boldsymbol{V}_- = \boldsymbol{V}_-\boldsymbol{T}_-$ holds, which we apply on $|M\rangle$ using (3.65) and (3.66):

$$\begin{aligned}
&\boldsymbol{T}_-\boldsymbol{V}_-\left|\tfrac{p}{2},\tfrac{p}{2}\right\rangle_V = \boldsymbol{V}_-\boldsymbol{T}_-\left|\tfrac{p}{2},\tfrac{p}{2}\right\rangle_T \\
&\boldsymbol{T}_-\sqrt{p}\left|\tfrac{p}{2},\tfrac{p-2}{2}\right\rangle_{V,(B)} = \boldsymbol{V}_-\sqrt{p}\left|\tfrac{p}{2},\tfrac{p-2}{2}\right\rangle_{T,(A)} \\
&\sqrt{p}\,\boldsymbol{T}_-\left|\tfrac{p-1}{2},\tfrac{p-1}{2}\right\rangle_{T,(B)} = \sqrt{p}\,\boldsymbol{V}_-\left|\tfrac{p-1}{2},\tfrac{p-1}{2}\right\rangle_{V,(A)} \\
&\sqrt{p\,(p-1)}\left|\tfrac{p-1}{2},\tfrac{p-3}{2}\right\rangle_{T,(N)} = \sqrt{p\,(p-1)}\left|\tfrac{p-1}{2},\tfrac{p-3}{2}\right\rangle_{V,(N)}.
\end{aligned} \tag{4.50}$$

If the state $|N\rangle$ would have double multiplicity, analogously to (4.37) the relation

$$\left|\frac{p-1}{2},\frac{p-3}{2}\right\rangle_{T(N)} = a_1\left|\frac{p-1}{2},\frac{p-3}{2}\right\rangle_{V(N)} + a_2\left|\frac{p-3}{2},\frac{p-3}{2}\right\rangle_{V(N)} \tag{4.51}$$

would be true. We replace the left-hand state by means of (4.50) and obtain

$$(1-a_1)\left|\frac{p-1}{2},\frac{p-3}{2}\right\rangle_{V(N)} = a_2\left|\frac{p-3}{2},\frac{p-3}{2}\right\rangle_{V(N)}. \tag{4.52}$$

But these orthogonal states cannot be proportional to one another. That is why the assumption of double multiplicity of $|N\rangle$ is wrong and $|N\rangle$ is unique like $|M\rangle$. As in the hexagonal case, the entire inner shell has uniform multiplicity. In the same way, one shows that all states in the triangle have multiplicity one. Analogously, if the multiplet has $q > 0$, the multiplicity of the inner triangle is uniformly $q+1$.

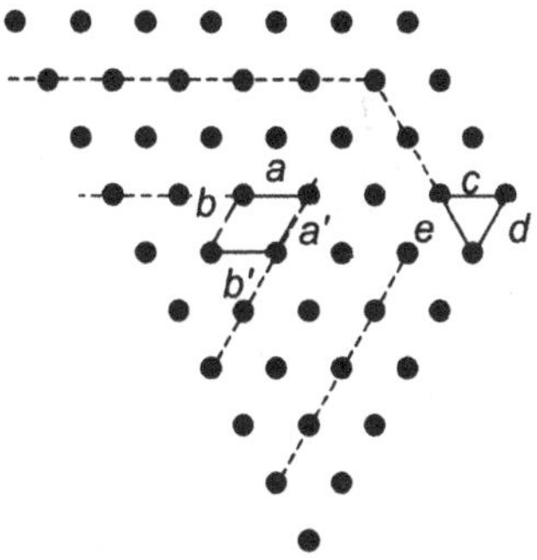

Figure 4.5.6. "short paths" in $su(3)$-multiplets.

We sum up these results as a simple recipe. Starting from a vertex of a shell we move inward on so-called "short paths" consisting of one or two unity steps. If there are two symmetric paths (for instance a, b and a', b' in Figure 4.5.6) the multiplicity is not raised. If there are two asymmetric "short paths" (for example c and de in Figure 4.5.6) the multiplicity is increased by 1.

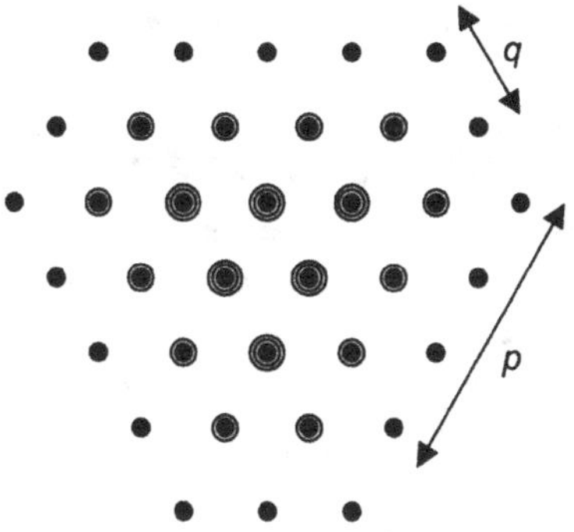

Figure 4.5.7. Multiplicities of the multiplet M(p,q).

In Figure 4.5.7 the additional (more than one) multiplicities are drawn as circles.

4.6 Dimension of the $su(3)$-multiplet

The sum of all multiplicities of all lattice points is named dimension d of the multiplet. Figure 4.5.7 shows that the boundary shell has $3(p+q)$ states. The next inner shell contains $3(p-1+q-1)$ doubly occupied points which yield $2 \cdot 3(p+q-2)$ partial states. On the nth inner shell (counted inward) are $(n+1) \cdot 3(p+q-2n)$ partial states. For $q \leq p$ we limit the index n at $q-1$. This value marks the last hexagonal shell. It has the multiplicity q. The shells with $0 \leq n \leq q-1$ contain together

$$\sum_{n=0}^{q-1}(n+1)3\,(p+q-2n) = -3\cdot 2\sum_{n=0}^{q-1} n^2 + 3\,(p+q-2)\sum_{n=0}^{q-1} n + 3\,(p+q)\,q \quad (4.53)$$

partial states. Still, we have to take into account an inner triangular or singular area of the multiplet with multiplicity $q+1$. Starting with the lowest point in the triangle in Figure 4.6.1, we count

$$\sum_{m=1}^{p-q+1} m = \frac{1}{2}\,(p-q+1)\,(p-q+2) \quad (4.54)$$

lattice points. The dimension d of the multiplet amounts to

$$\begin{aligned} d\,(p,q) &= \frac{1}{2}\,(p-q+1)\,(p-q+2)\,(q+1) \\ &\quad -6\sum_{n=0}^{q-1} n^2 + 3\,(p+q-2)\sum_{n=0}^{q-1} n + 3\,(p+q)\,q. \end{aligned} \quad (4.55)$$

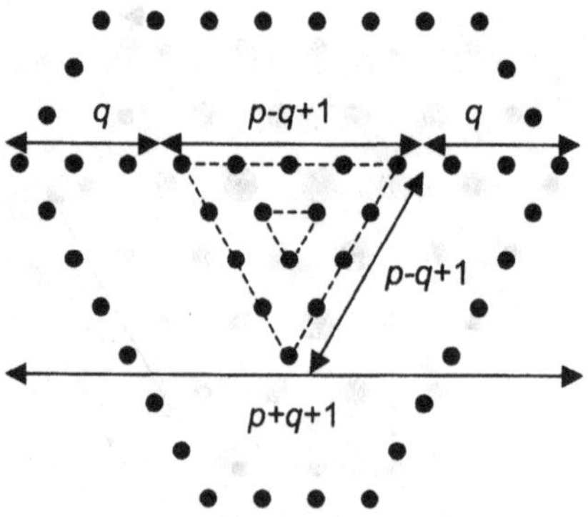

Figure 4.6.1. The triangular area.

Making use of $\sum_{n=0}^{q-1} n^2 = \frac{q}{6}\,(q-1)\,(2q-1)$ and $\sum_{n=0}^{q-1} n = \frac{q}{2}\,(q-1)$ finally we obtain

$$d\,(p,q) = \frac{1}{2}\,(p+1)\,(q+1)\,(p+q+2)\,. \quad (4.56)$$

The dimension d is symmetric in p and q, which follows also from the fact that interchanging both parameters turns the multiplet diagram by 60^0. For the simplest multiplets the dimensions are given in Table 4.6.1, which is calculated by means of (4.56).

Table 4.6.1. Dimensions of the lowest $su(3)$-multiplets.

p	q	$d(p,q)$	p	q	$d(p,q)$
0	0	1	3	0	10
1	0	3	3	1	32
1	1	8	3	2	42
2	0	6	3	3	64
2	1	15			
2	2	27			

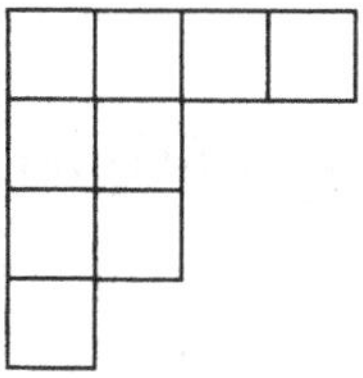

Figure 4.6.2. Young diagram.

On this occasion, we acquaint ourselves with the mathematical-geometrical method of **Young tableaux**. Such a tableau consists of rows of boxes. Every lower row is as long or shorter than the upper one. Integers from 1 up to a maximum are put — repetitions are permitted — in the boxes in such a way that to the right the values do not decrease and downward in a column they increase. This scheme is named *standard arrangement.*

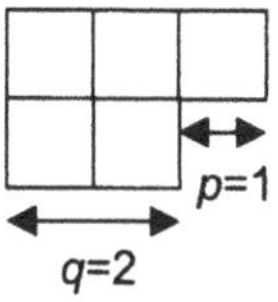

Figure 4.6.3. A $su(3)$-Young diagram.

For $su(3)$-multiplets we draw two lines of boxes. The number of boxes in the first line which jut over the second line is chosen to be p, and the second line has q boxes. The integers 1, 2 and 3 are available to be put in, eventually repeatedly.

A **theorem**, which we do not prove, is that the number d of standard arrangements of this Young tableau is equal to the dimension of the multiplet $M(p,q)$. We write down the arrangements for $p = 1$ and $q = 2$ (corresponding to Figure 4.6.3):

$$\begin{array}{ccc}1&1&1\\2&2&\end{array}\quad\begin{array}{ccc}1&1&1\\2&3&\end{array}\quad\begin{array}{ccc}1&1&1\\3&3&\end{array}\quad\begin{array}{ccc}1&1&2\\2&2&\end{array}\quad\begin{array}{ccc}1&1&2\\2&3&\end{array}$$

$$\begin{array}{ccc}1&1&2\\3&3&\end{array}\quad\begin{array}{ccc}1&1&3\\2&2&\end{array}\quad\begin{array}{ccc}1&1&3\\2&3&\end{array}\quad\begin{array}{ccc}1&1&3\\3&3&\end{array}\quad\begin{array}{ccc}1&2&2\\2&3&\end{array}$$

$$\begin{array}{ccc}1&2&2\\3&3&\end{array}\quad\begin{array}{ccc}1&2&3\\2&3&\end{array}\quad\begin{array}{ccc}1&2&3\\3&3&\end{array}\quad\begin{array}{ccc}2&2&2\\3&3&\end{array}\quad\begin{array}{ccc}2&2&3\\3&3&\end{array}$$

One counts 15 arrangements. According to the given theorem the dimension $d(1,2)$ is 15, which is in accordance with Table 4.6.1. In Sections 4.14, 5.5 and 6.2 we again will find Young tableaux.

4.7 The smallest $su(3)$-multiplets

We show the simplest $su(3)$-multiplets $M_d\,(p,q)$ with $p,q \leq 3$. In this symbol the dimension is put as a subscript character. The coordinate axes are placed in the centre of gravity in every figure. The numerical details on the y-axis (integers or thirds) will be looked at in Section 4.9.

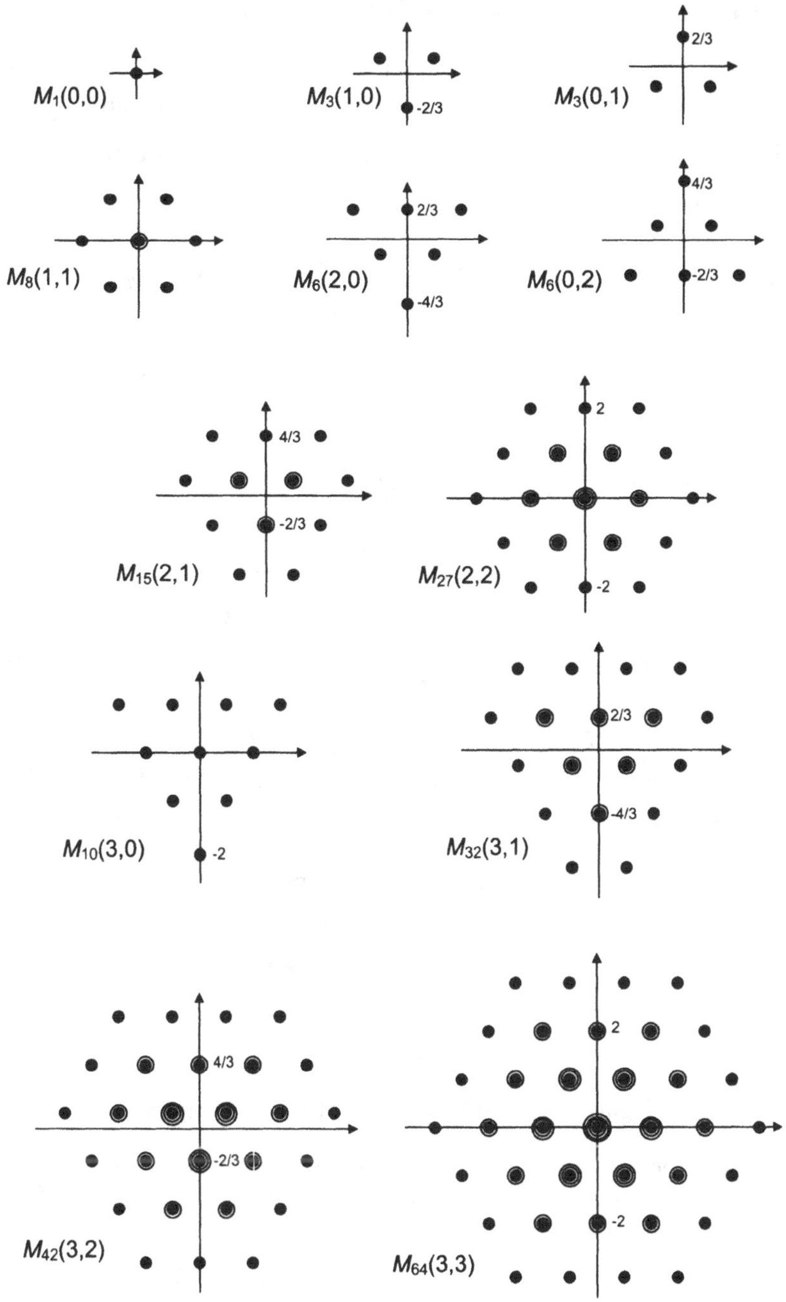

Figure 4.7.1.The simplest $su(3)$-multiplets.

4.8 The fundamental multiplet of $su(3)$

The operators $\boldsymbol{\lambda}_i$ of $su(3)$ were introduced at the beginning of Section 4.2. If they act on a function of the fundamental multiplet $\{\psi_1, \psi_2, \psi_3\}$, according to (2.19) and (2.22), they meet $\boldsymbol{\lambda}_j \psi_k = \sum_{l=1}^{3} (\lambda_j)_{lk} \psi_l$, and, taking into account (4.9) the following relations hold:

$$
\begin{aligned}
\boldsymbol{T}_+ \psi_k &= \frac{1}{2}(\boldsymbol{\lambda}_1 + \mathrm{i}\boldsymbol{\lambda}_2)\,\psi_k = \sum_{l=1}^{3} \frac{1}{2}\left(\underline{\underline{\lambda}}_1 + \mathrm{i}\underline{\underline{\lambda}}_2\right)_{lk} \psi_l \\
&= \left[(\psi_1, \psi_2, \psi_3)\begin{pmatrix} 0 & 1 & 0 \\ 0 & 0 & 0 \\ 0 & 0 & 0 \end{pmatrix}\right]_k = \psi_1\, \delta_{k2}, \\
\boldsymbol{T}_- \psi_k &= \sum_{l=1}^{3} \frac{1}{2}\left(\underline{\underline{\lambda}}_1 - \mathrm{i}\underline{\underline{\lambda}}_2\right)_{lk} \psi_l \qquad (4.57)\\
&= \left[(\psi_1, \psi_2, \psi_3)\begin{pmatrix} 0 & 0 & 0 \\ 1 & 0 & 0 \\ 0 & 0 & 0 \end{pmatrix}\right]_k = \psi_2\, \delta_{k1}
\end{aligned}
$$

and so forth. We have inserted the matrices $\underline{\underline{\lambda}}_i$ from (4.1) and (4.2). The results read

$$
\begin{array}{llll}
\boldsymbol{T}_+\psi_2 = \psi_1, & \boldsymbol{T}_-\psi_1 = \psi_2, & \boldsymbol{T}_3\psi_1 = \frac{1}{2}\psi_1, & \boldsymbol{T}_3\psi_2 = -\frac{1}{2}\psi_2, \\
\boldsymbol{U}_+\psi_3 = \psi_2, & \boldsymbol{U}_-\psi_2 = \psi_3, & \boldsymbol{U}_3\psi_2 = \frac{1}{2}\psi_2, & \boldsymbol{U}_3\psi_3 = -\frac{1}{2}\psi_3, \\
\boldsymbol{V}_+\psi_3 = \psi_1, & \boldsymbol{V}_-\psi_1 = \psi_3, & \boldsymbol{V}_3\psi_1 = \frac{1}{2}\psi_1, & \boldsymbol{V}_3\psi_3 = -\frac{1}{2}\psi_3.
\end{array}
\qquad (4.58)
$$

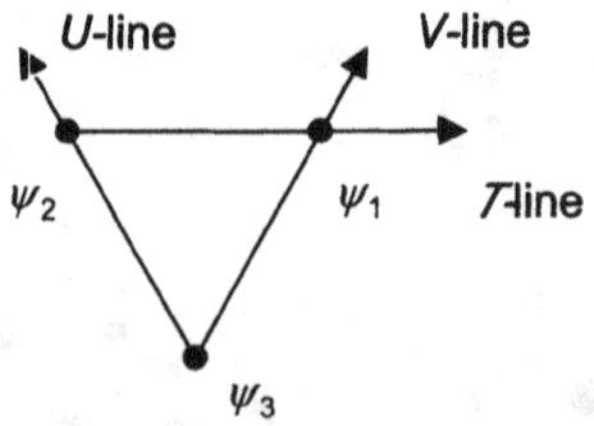

Figure 4.8.1. The fundamental $su(3)$-multiplet.

Other applications of these operators on these functions yield zero. We maintain that the fundamental multiplet $\{\psi_1, \psi_2, \psi_3\}$ is identical with $M_3\,(1,0)$ if the following assignments are chosen (see Figure 4.8.1):

$$
\begin{aligned}
\psi_1 &= \left|\tfrac{1}{2}, \tfrac{1}{2}\right\rangle_T &&= |0,0\rangle_U &&= \left|\tfrac{1}{2}, \tfrac{1}{2}\right\rangle_V, \\
\psi_2 &= \left|\tfrac{1}{2}, -\tfrac{1}{2}\right\rangle_T &&= \left|\tfrac{1}{2}, \tfrac{1}{2}\right\rangle_U &&= |0,0\rangle_V, \qquad (4.59)\\
\psi_3 &= |0,0\rangle_T &&= \left|\tfrac{1}{2}, -\tfrac{1}{2}\right\rangle_U &&= \left|\tfrac{1}{2}, -\tfrac{1}{2}\right\rangle_V.
\end{aligned}
$$

The $su(2)$-operators $\boldsymbol{T}_+$, $\boldsymbol{T}_-$, $\boldsymbol{T}_3$ and the operators $\boldsymbol{U}_+$, $\boldsymbol{U}_-$, $\boldsymbol{U}_3$ and $\boldsymbol{V}_+$, $\boldsymbol{V}_-$, $\boldsymbol{V}_3$ act fully equivalent to the operators $\boldsymbol{J}_+$, $\boldsymbol{J}_-$, $\boldsymbol{J}_3$, (3.41) (see the comment preceding equations (4.41)). Therefore, we can use the relations (3.60) and (3.67) in order to recalculate the expressions (4.58). The results coincide entirely with (4.58).

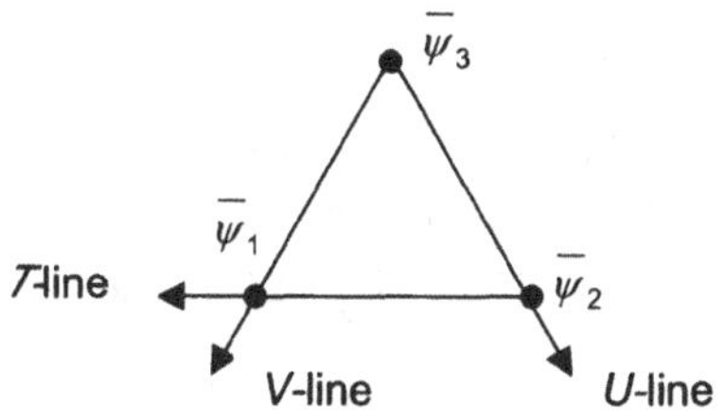

Figure 4.8.2. The fundamental antimultiplet.

The multiplet $M_3(0,1)$ of Figure 4.8.2 is also a fundamental multiplet. The T-, U- and V-lines are reversed (see the remark at the end of Section 4.3). In this way, the relations (4.59) and (4.58) are also satisfied.

In particle physics the multiplet $M_3(1,0)$ is denoted by [3] and describes the quark states whereas $M_3(0,1) \equiv [\bar{3}]$ stands for the antiquark states.

4.9 The hypercharge *Y*

In order to specify the status of a $su(3)$-multiplet we used the T_3-, U_3- and V_3-values in a star-shaped coordinate system. Of course, two coordinates are sufficient for a two-dimensional system. In practice, the T-axis and a perpendicular Y-axis are taken. In particle physics the Y-values are named hypercharge. We claim that the coordinate Y is the eigenvalue of the operator

$$\boldsymbol{Y} = \frac{1}{\sqrt{3}}\boldsymbol{\lambda}_8 = \frac{2}{3}(\boldsymbol{U}_3 + \boldsymbol{V}_3) = \frac{2}{3}(2\boldsymbol{U}_3 + \boldsymbol{T}_3) = \frac{2}{3}(2\boldsymbol{V}_3 - \boldsymbol{T}_3). \tag{4.60}$$

Here the relations (4.11) and (4.13) have been applied. From (4.60) and Table 4.2.1 it follows

$$[\boldsymbol{Y}, \boldsymbol{T}_3] = [\boldsymbol{Y}, \boldsymbol{U}_3] = [\boldsymbol{Y}, \boldsymbol{V}_3] = 0. \tag{4.61}$$

Therefore, the state $|T_3U_3V_3\rangle$ is also eigenstate of the operator $\mathbf{Y}$, i.e.,

$$\boldsymbol{Y}\,|T_3U_3V_3\rangle = Y\,|T_3U_3V_3\rangle \quad \text{and} \quad |T_3U_3V_3\rangle \equiv |T_3Y\rangle\,. \tag{4.62}$$

By means of Table 4.2.1, we can write

$$\begin{aligned}&[\boldsymbol{Y}, \boldsymbol{T}_\pm] = \tfrac{2}{3}[\boldsymbol{U}_3, \boldsymbol{T}_\pm] + \tfrac{2}{3}[\boldsymbol{V}_3, \boldsymbol{T}_\pm] = \tfrac{2}{3}(\mp\boldsymbol{T}_\pm \pm \boldsymbol{T}_\pm) = 0 \quad \text{and}\\ &\boldsymbol{Y}(\boldsymbol{T}_\pm\,|T_3Y\rangle) = \boldsymbol{T}_\pm\boldsymbol{Y}\,|T_3Y\rangle = Y(\boldsymbol{T}_\pm\,|T_3Y\rangle)\,.\end{aligned} \tag{4.63}$$

That is, the state $T_\pm |T_3 Y\rangle$,which differs in T_3 by $\pm$ 1 from the state $|T_3 Y\rangle$, has the same Y as $|T_3 Y\rangle$. This corroborates that the Y-axis is orthogonal to the T-line. Similarly one derives $[\boldsymbol{Y}, \boldsymbol{U}_\pm] = \pm \boldsymbol{U}_\pm$ which yields

$$\begin{aligned} \boldsymbol{Y}(\boldsymbol{U}_\pm |T_3 Y\rangle) &= \boldsymbol{U}_\pm \boldsymbol{Y} |T_3 Y\rangle \pm \boldsymbol{U}_\pm |T_3 Y\rangle \\ &= (Y \pm 1)(\boldsymbol{U}_\pm |T_3 Y\rangle) \equiv (Y \pm 1)(\boldsymbol{U}_\pm |T_3 \boldsymbol{U}_3 \boldsymbol{V}_3\rangle). \quad (4.64) \end{aligned}$$

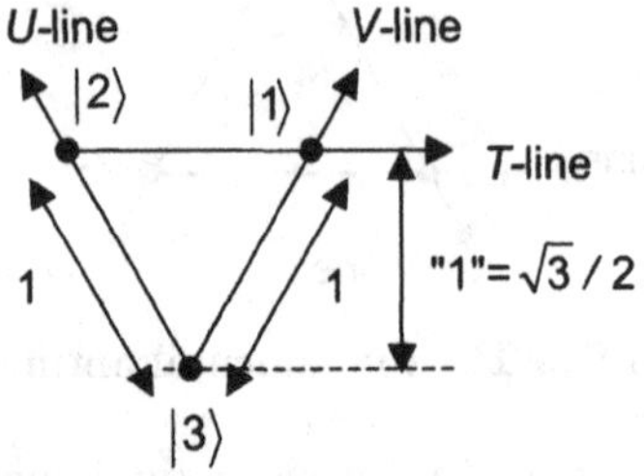

Figure 4.9.1. Step 1 in the directions U, V and Y.

This means that $\boldsymbol{U}_\pm$ not only rises and lowers U_3 by 1 but also Y in the same sense. Analogously one shows that Y varies alike to V_3. Figure 4.9.1 shows this behaviour. Points 1 and 2 have both the distance 1 from 3. On the other hand, the distance (Y-value) of 3 from the T-line is also 1, which is geometrically impossible. We circumvent the inconsistency by taking the Y-values in units of $\sqrt{3}/2$. The geometrical Y-distance $\sqrt{3}/2$ is interpreted as a formal value "1".

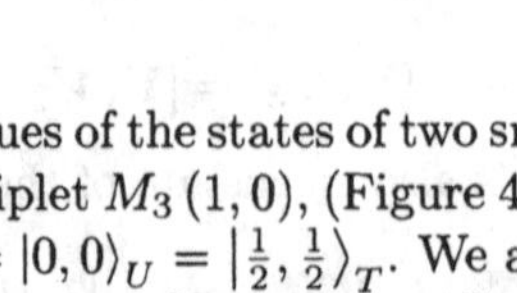

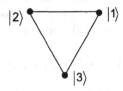

Figure 4.9.1a. The multiplet $M_3(1{,}0)$.

We now determine the Y-values of the states of two small multiplets and of their coordinate origins. In the multiplet $M_3(1,0)$, (Figure 4.9.1a), the state $|1\rangle$, according to (4.59), is written as $|1\rangle = |0,0\rangle_U = \left|\frac{1}{2}, \frac{1}{2}\right\rangle_T$. We apply the operator $\boldsymbol{Y}$, according to (4.60), to this state: $\boldsymbol{Y}|1\rangle = \frac{4}{3}\boldsymbol{U}_3 |0,0\rangle_U + \frac{2}{3}\boldsymbol{T}_3 \left|\frac{1}{2}, \frac{1}{2}\right\rangle_T = \frac{1}{3}\left|\frac{1}{2}, \frac{1}{2}\right\rangle_T = \frac{1}{3}|1\rangle$. The eigenvalue of the $\boldsymbol{Y}$-operator is 1/3 in $|1\rangle$, which holds also for $|2\rangle$. For $|3\rangle = \left|\frac{1}{2}, -\frac{1}{2}\right\rangle_U = |0,0\rangle_T$ we obtain $\boldsymbol{Y}|3\rangle = \frac{4}{3}\boldsymbol{U}_3 \left|\frac{1}{2}, -\frac{1}{2}\right\rangle_U + \frac{2}{3}\boldsymbol{T}_3 |0,0\rangle_T = -\frac{2}{3}\left|\frac{1}{2}, \frac{1}{2}\right\rangle_U$ $= -\frac{2}{3}|3\rangle$ as indicated in Figure 4.7.1. Clearly the origin of the coordinate system is 1/3 below the T-line and the height of the triangle amounts to "1".

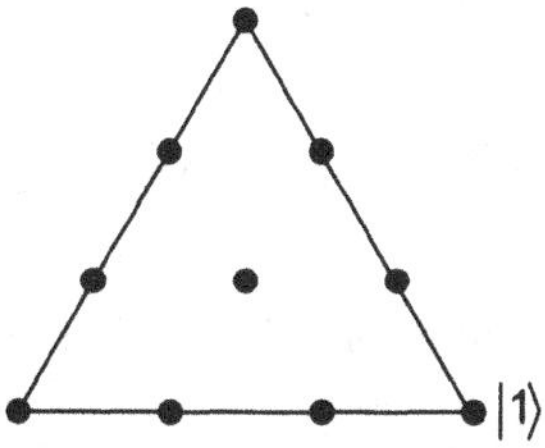

Figure 4.9.2. The multiplet $M_{10}(0,3)$.

In the multiplet $M_{10}(0,3)$ we have $|1\rangle = \left|\frac{3}{2},-\frac{3}{2}\right\rangle_U = |0,0\rangle_V$ and

$$\boldsymbol{Y}\,|1\rangle = \frac{2}{3}\boldsymbol{U}_3\left|\frac{3}{2},-\frac{3}{2}\right\rangle_U + \frac{2}{3}\boldsymbol{V}_3\,|0,0\rangle_V = -1\left|\frac{3}{2},-\frac{3}{2}\right\rangle_U = -1\,|1\rangle\,.$$

Therefore, the origin lies one unit above the meshpoint $|1\rangle$.

Obviously, the origin of the TY-coordinate system lies on the intersection point of the symmetry axes s and s' (see Figure 4.9.3), because there the eigenvalues U_3 and V_3 vanish, i.e.,

$$\boldsymbol{Y}\,|T_3U_3V_3\rangle = \frac{2}{3}\,(U_3+V_3)\,|T_3U_3V_3\rangle = 0. \tag{4.65}$$

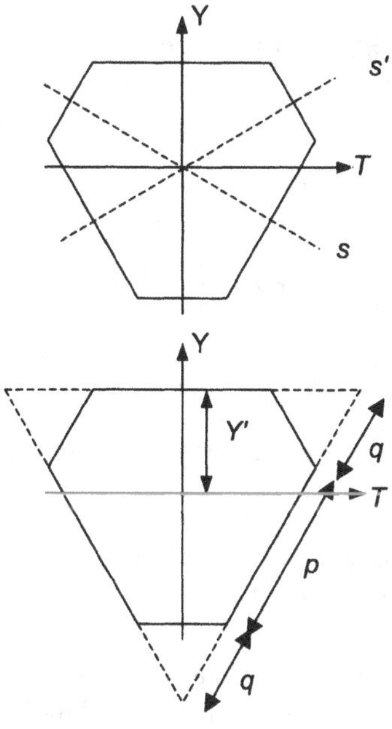

Figure 4.9.3. Coordinate system TY.

The value Y' in Figure 4.9.3 is for planimetrical reasons (origin = center of gravity of the figure)

$$Y' = \frac{1}{3} \cdot \frac{\sqrt{3}}{2}(p+2q) = \frac{\sqrt{3}}{2}\left(q + \frac{p-q}{3}\right). \tag{4.66}$$

Since p and q are integers, the expression $\frac{p-q}{3}$ amounts to an integer number plus 0, 1/3 or 2/3. Therefore, the points on top of the multiplet can have the Y-coordinates integer, integer + 1/3 or integer + 2/3 (measured in units of $\sqrt{3}/2$). This is true for all the states of the multiplet because they differ by integers in Y. One says that the multiplet has the

$$\textit{triality}\ \frac{\tau}{3} \quad (\tau = 0, 1, 2) \tag{4.67}$$

if its coordinates are an integer plus $\frac{\tau}{3}$.

4.10 Irreducible representations of the $su(3)$ algebra

Due to (3.32) and (1.49) we have

$$\boldsymbol{e}_i\psi_k = \sum_{l=1}^{d} \Gamma(\boldsymbol{e}_i)_{lk}\psi_l = \sum_{l=1}^{d}(-\frac{1}{2}\mathrm{i})\Gamma(\boldsymbol{\lambda}_i)_{lk}\psi_l. \tag{4.68}$$

Therefore, the matrices $-\frac{1}{2}\mathrm{i}\underline{\underline{\Gamma}}(\boldsymbol{\lambda}_i)$ are a representation of the Lie algebra $su(3)$. Equation (4.68) can be written as

$$\boldsymbol{\lambda}_i\psi_k = \sum_{l=1}^{d} \Gamma(\boldsymbol{\lambda}_i)_{lk}\psi_l.$$

According to (1.36) the matrix elements of $\underline{\underline{\Gamma}}(\boldsymbol{\lambda}_i)$ read

$$\Gamma(\boldsymbol{\lambda}_i)^{(p,q)}_{mk} = \left\langle \psi^{(p,q)}_m \,\middle|\, \boldsymbol{\lambda}_i\, \psi^{(p,q)}_k \right\rangle. \tag{4.69}$$

By means of (4.9) and (4.11) we can write the operators $\boldsymbol{\lambda}_i$ this way:

$$\begin{aligned}
&\boldsymbol{\lambda}_1 = \boldsymbol{T}_+ + \boldsymbol{T}_-, && \boldsymbol{\lambda}_2 = -\mathrm{i}(\boldsymbol{T}_+ - \boldsymbol{T}_-), && \boldsymbol{\lambda}_3 = 2\boldsymbol{T}_3,\\
&\boldsymbol{\lambda}_4 = \boldsymbol{V}_+ + \boldsymbol{V}_-, && \boldsymbol{\lambda}_5 = -\mathrm{i}(\boldsymbol{V}_+ - \boldsymbol{V}_-), && \\
&\boldsymbol{\lambda}_6 = \boldsymbol{U}_+ + \boldsymbol{U}_-, && \boldsymbol{\lambda}_7 = -\mathrm{i}(\boldsymbol{U}_+ - \boldsymbol{U}_-), && \boldsymbol{\lambda}_8 = \tfrac{2}{\sqrt{3}}(\boldsymbol{U}_3 + \boldsymbol{V}_3).
\end{aligned} \tag{4.70}$$

For the moment, we do not make use of the operator $\mathbf{Y}$. Given a $su(3)$-multiplet, the matrix elements (4.69) can be calculated using the $su(2)$-rules (3.60), (3.65) and (3.66) for the $\boldsymbol{T}$-, $\boldsymbol{U}$-, and $\boldsymbol{V}$-operators.

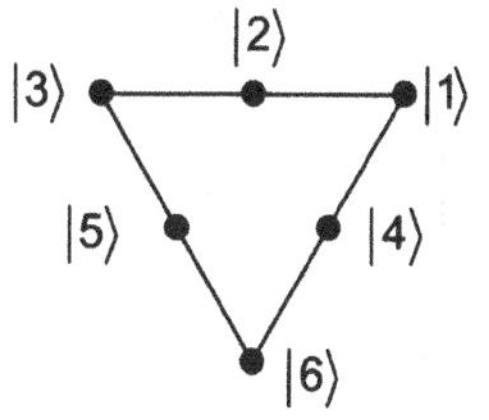

Figure 4.10.1. Multiplet $M_6(2,0)$.

For **example**, we determine **the $su(3)$ representation of the multiplet** $M_6(2,0)$ (Figure 4.10.1). The states $|1\rangle$, $|2\rangle$, ..., $|6\rangle$ are characterised by the following T_3, U_3 and V_3 values:

state	T_3	U_3	V_3
1	1	0	1
2	0	1/2	1/2
3	−1	1	0
4	1/2	−1/2	0
5	−1/2	0	−1/2
6	0	−1	−1.

If the operators $\boldsymbol{T}_\pm$, $\boldsymbol{U}_\pm$ or $\boldsymbol{V}_\pm$ act on triplet $su(2)$-states, a factor $\sqrt{2}$ appears according to (3.67), and doublet states have only the factor 1. We obtain

$$\begin{array}{lll} & \boldsymbol{T}_- |1\rangle = \sqrt{2}\,|2\rangle\,, & \boldsymbol{T}_3\,|1\rangle = |1\rangle\,, \\ \boldsymbol{T}_+ |2\rangle = \sqrt{2}\,|1\rangle\,, & \boldsymbol{T}_- |2\rangle = \sqrt{2}\,|3\rangle\,, & \\ \boldsymbol{T}_+ |3\rangle = \sqrt{2}\,|2\rangle\,, & & \boldsymbol{T}_3\,|3\rangle = -\,|3\rangle\,, \\ & \boldsymbol{T}_- |4\rangle = |5\rangle\,, & \boldsymbol{T}_3\,|4\rangle = \frac{1}{2}\,|4\rangle\,, \\ \boldsymbol{T}_+ |5\rangle = |4\rangle\,, & & \boldsymbol{T}_3\,|5\rangle = -\frac{1}{2}\,|5\rangle\,. \end{array} \tag{4.71}$$

The remaining expressions with **T**-operators vanish. For the **U**- and the **V**-operators analogous relations hold. Therefore, due to (4.69) and (4.70), for $p = 2$ and $q = 0$ we have, for example,

$$\Gamma\,(\boldsymbol{\lambda}_1)^{(2,0)}_{mk} = \langle\psi_m \mid \boldsymbol{T}_+\psi_k\rangle + \langle\psi_m \mid \boldsymbol{T}_-\psi_k\rangle\,, \tag{4.72}$$

which yields for $m = 1$ and $k = 2$:

$$\Gamma\,(\boldsymbol{\lambda}_1)^{(2,0)}_{1,2} = \sqrt{2} + 0. \tag{4.73}$$

With the rest of the matrix elements and of the operators $\boldsymbol{\lambda}_i$ one has to proceed similarly and obtains the 6×6-matrices $\underline{\underline{\Gamma}}\,(\boldsymbol{\lambda}_i)^{(2,0)}$ representing $su(3)$ like this:

$$\underline{\underline{\Gamma}}(\boldsymbol{\lambda}_1)^{(2,0)} = \left(\begin{array}{cccccc} 0 & \sqrt{2} & 0 & & & \\ \sqrt{2} & 0 & \sqrt{2} & & 0 & \\ 0 & \sqrt{2} & 0 & & & \\ & & & 0 & 1 & 0 \\ & 0 & & 1 & 0 & 0 \\ & & & 0 & 0 & 0 \end{array}\right),$$

$$\underline{\underline{\Gamma}}(\boldsymbol{\lambda}_2)^{(2,0)} = \mathrm{i} \left(\begin{array}{cccccc} 0 & \sqrt{2} & 0 & & & \\ \sqrt{2} & 0 & \sqrt{2} & & 0 & \\ 0 & \sqrt{2} & 0 & & & \\ & & & 0 & -1 & 0 \\ & 0 & & 1 & 0 & 0 \\ & & & 0 & 0 & 0 \end{array}\right),$$

$$\underline{\underline{\Gamma}}(\boldsymbol{\lambda}_3)^{(2,0)} = \left(\begin{array}{cccccc} 2 & 0 & 0 & & & \\ 0 & 0 & 0 & & 0 & \\ 0 & 0 & -2 & & & \\ & & & 1 & 0 & 0 \\ & 0 & & 0 & -1 & 0 \\ & & & 0 & 0 & 0 \end{array}\right),$$

$$\underline{\underline{\Gamma}}(\boldsymbol{\lambda}_4)^{(2,0)} = \left(\begin{array}{cccccc} & & & \sqrt{2} & 0 & 0 \\ & 0 & & 0 & 1 & 0 \\ & & & 0 & 0 & 0 \\ \sqrt{2} & 0 & 0 & 0 & 0 & \sqrt{2} \\ 0 & 1 & 0 & 0 & 0 & 0 \\ 0 & 0 & 0 & \sqrt{2} & 0 & 0 \end{array}\right),$$

$$\underline{\underline{\Gamma}}(\boldsymbol{\lambda}_5)^{(2,0)} = \mathrm{i} \left(\begin{array}{cccccc} & & & -\sqrt{2} & 0 & 0 \\ & 0 & & 0 & -1 & 0 \\ & & & 0 & 0 & 0 \\ \sqrt{2} & 0 & 0 & 0 & 0 & \sqrt{2} \\ 0 & 1 & 0 & 0 & 0 & 0 \\ 0 & 0 & 0 & \sqrt{2} & 0 & 0 \end{array}\right),$$

$$\underline{\underline{\Gamma}}(\boldsymbol{\lambda}_6)^{(2,0)} = \left(\begin{array}{cccccc} & & & 0 & 0 & 0 \\ & 0 & & 1 & 0 & 0 \\ & & & 0 & \sqrt{2} & 0 \\ 0 & 1 & 0 & 0 & 0 & 0 \\ 0 & 0 & \sqrt{2} & 0 & 0 & \sqrt{2} \\ 0 & 0 & 0 & 0 & \sqrt{2} & 0 \end{array}\right),$$

$$
\underline{\underline{\Gamma}}(\boldsymbol{\lambda}_7)^{(2,0)} = \mathrm{i}\begin{pmatrix} 0 & 0 & 0 & 0 & 0 & 0 \\ 0 & 0 & 0 & -1 & 0 & 0 \\ 0 & 0 & 0 & 0 & -\sqrt{2} & 0 \\ 0 & 1 & 0 & 0 & 0 & 0 \\ 0 & 0 & \sqrt{2} & 0 & 0 & -\sqrt{2} \\ 0 & 0 & 0 & 0 & \sqrt{2} & 0 \end{pmatrix},
$$

$$
\underline{\underline{\Gamma}}(\boldsymbol{\lambda}_8)^{(2,0)} = \begin{pmatrix} 1 & 0 & 0 & 0 & 0 & 0 \\ 0 & 1 & 0 & 0 & 0 & 0 \\ 0 & 0 & 1 & 0 & 0 & 0 \\ 0 & 0 & 0 & -\frac{1}{2} & 0 & 0 \\ 0 & 0 & 0 & 0 & -\frac{1}{2} & 0 \\ 0 & 0 & 0 & 0 & 0 & -2 \end{pmatrix}. \tag{4.74}
$$

By numerical calculation one makes sure that the relation

$$
\left[\underline{\underline{\Gamma}}(\boldsymbol{\lambda}_i)^{(2,0)}, \underline{\underline{\Gamma}}(\boldsymbol{\lambda}_j)^{(2,0)}\right] = 2\mathrm{i}C_{ijk}\underline{\underline{\Gamma}}(\boldsymbol{\lambda}_k)^{(2,0)} \tag{4.75}
$$

is satisfied, which we expect from considerations in Section 1.4 and from equation (2.26). The C_{ijk}'s are given in Table 4.1.1.

If one works on the simple "quark" multiplet $M_3(1,0)$, proceeding in the same way one produces a representation of $su(3)$ with 3×3-matrices. If the state numbers and the coordinate system are chosen as in Figure 4.8.1, the numerical calculation reveals that the representation matrices $\underline{\underline{\Gamma}}(\boldsymbol{\lambda}_i)^{(1,0)}$ coincide with the corresponding matrices $\underline{\underline{\lambda}}_i$, which is a trivial consequence of the definition (2.19).

In order to obtain a representation from the "antiquark"-multiplet $M_3(0,1)$, we have to take the "anti" coordinate system of Figure 4.3.3 or of Figure 4.8.2. The resulting matrices $\underline{\underline{\Gamma}}(\boldsymbol{\lambda}_i)^{(0,1)}$ coincide again with $\underline{\underline{\lambda}}_i$.

4.11 Casimir operators

In the $su(2)$-algebra the operator $\boldsymbol{J}^2$ plays an important role. On the one hand, it commutes with the elements $\boldsymbol{J}_+, \boldsymbol{J}_-, \boldsymbol{J}_3$ (see (3.43)) and therefore with $\boldsymbol{e}_1$, $\boldsymbol{e}_2$, $\boldsymbol{e}_3$, i.e., with all elements of $su(2)$, and, on the other hand, the functions of the $su(2)$-multiplets are eigenfunctions of $\boldsymbol{J}^2$, according to (3.59). The eigenvalue has the form $j\,(j+1)$, which contains the integer or half integer quantum number j. It is characteristic for the multiplet under consideration.

Operators which commute with all elements $\boldsymbol{e}_i$ of a Lie algebra are named *Casimir operators.* Because of $\boldsymbol{e}_i = -\frac{1}{2}\mathrm{i}\boldsymbol{\lambda}_i$, they commute also with the generators $\boldsymbol{\lambda}_i$. Generally, one can show that the states of a multiplet accompanying the algebra are eigenfunctions of the Casimir operator and have a uniform eigenvalue (part of the Theorem of Racah). Furthermore, it can be proven that the number of Casimir operators attributed to a Lie algebra is identical with the rank l of the algebra (see at the end of Section 4.1). Generally, it can be shown that the algebra $su(N)$ has rank $N-1$. Therefore, in $su(2)$ the operator $\boldsymbol{J}^2$ is the unique Casimir operator and $su(3)$ has two such operators.

In analogy with $su(2)$, all functions of a multiplet are eigenfunctions of every Casimir operator. This simultaneous property of the Casimir operators $\boldsymbol{C}_1$, $\boldsymbol{C}_2, \ldots, \boldsymbol{C}_j$ is based on the fact that they are constructed as combinations of quadratic or higher expressions of the generators (see equation (4.77) below). As a consequence, a Casimir operator $\boldsymbol{C}_j$ commutes with every partial operator $\boldsymbol{\lambda}_i$ of another Casimir operator $\boldsymbol{C}_{j'}$ and therefore with the entire operator, i.e.,

$$\left[\boldsymbol{C}_j, \boldsymbol{C}_j'\right] = 0, \tag{4.76}$$

which results — as we know — in simultaneous eigenvalue equations of these operators. They yield the eigenvalues $C_1, C_2, \ldots, C_l$. Practically, these values are formulated by l independent parameters $p_1, p_2, \ldots, p_l$, which characterise the multiplets instead of the C's. The integers p and q, which appeared in $su(3)$ at the end of Section 4.4, are parameters of this kind. In the next section we will see how the eigenvalue of $\boldsymbol{C}_1$ is composed out of them.

Casimir operators can be defined in diverse ways. Because the sum of two operators of this kind commutes also with the generators it is a Casimir operator as well. This holds also for a polynomial of such an operator or for a mixed polynomial of several Casimir operators.

We will show that the **quadratic Casimir operator** $\boldsymbol{C}_1$ of $su(N)$ has the following form:

$$\boldsymbol{C}_1 = -\frac{1}{4}\sum_{i=1}^{n}\boldsymbol{\lambda}_i^2. \tag{4.77}$$

The operators $-\frac{1}{2}\mathrm{i}\boldsymbol{\lambda}_i$ constitute the basis of the algebra, and n is its dimension. In $su(2)$ the expression (4.77) reads

$$\boldsymbol{C}_1^{su(2)} = -\left(\boldsymbol{J}_x^2 + \boldsymbol{J}_y^2 + \boldsymbol{J}_z^2\right) = -\boldsymbol{J}^2. \tag{4.78}$$

Of course, $+\boldsymbol{J}^2$ can be taken as well. We now prove (4.77) and check whether $\boldsymbol{C}_1$ commutes with an arbitrary element $\boldsymbol{\lambda}_k$.

$$\begin{aligned}4\left[\boldsymbol{C}_1, \boldsymbol{\lambda}_k\right] &= -\sum_{i=1}^{n}\left[\boldsymbol{\lambda}_i^2, \boldsymbol{\lambda}_k\right] = -\sum_{i=1}^{n}\left(\boldsymbol{\lambda}_i\left[\boldsymbol{\lambda}_i, \boldsymbol{\lambda}_k\right] + \left[\boldsymbol{\lambda}_i, \boldsymbol{\lambda}_k\right]\boldsymbol{\lambda}_i\right) \\ &= -2\mathrm{i}\sum_{i=1}^{n}\sum_{m=1}^{n}\left(C_{ikm}\boldsymbol{\lambda}_i\boldsymbol{\lambda}_m + C_{ikm}\boldsymbol{\lambda}_m\boldsymbol{\lambda}_i\right) = 0.\end{aligned} \tag{4.79}$$

In the last term we have replaced and interchanged the indices and made use of (2.30) like this:

$$\sum_{i,m=1}^{n} C_{ikm}\boldsymbol{\lambda}_m\boldsymbol{\lambda}_i = \sum_{m',i'=1}^{n} C_{m'ki'}\boldsymbol{\lambda}_{i'}\boldsymbol{\lambda}_{m'} = -\sum_{m',i'=1}^{n} C_{i'km'}\boldsymbol{\lambda}_{i'}\boldsymbol{\lambda}_{m'} = -\sum_{i,m=1}^{n} C_{ikm}\boldsymbol{\lambda}_i\boldsymbol{\lambda}_m. \tag{4.80}$$

It can be shown that the Casimir operator $\boldsymbol{C}_2$ contains the generators $\boldsymbol{\lambda}_i$ in third order. However no general constructing procedure is known. Anyhow, in $su(3)$ the expression

$$\boldsymbol{C}_2 = \boldsymbol{C}_1\left(2\boldsymbol{C}_1 - \frac{11}{6}\right) \tag{4.81}$$

can be given. It can be transformed in another one with the $\boldsymbol{\lambda}_i$'s in third order, which looks very similar to a transformed expression for $\boldsymbol{C}_1$ (Greiner, Müller, 1994, p. 197).

4.12 The eigenvalue of the Casimir operator C_1 in $su(3)$

If we change the sign in equation (4.77) the essential properties of the Casimir operator remain unchanged. This form is also very common and will be used here:

$$\boldsymbol{C}_1 = \frac{1}{4}\sum_{i=1}^{8}\boldsymbol{\lambda}_i^2 \tag{4.82}$$

(quadratic Casimir operator). By means of (4.70), we write $\frac{1}{4}\left(\boldsymbol{\lambda}_1^2+\boldsymbol{\lambda}_2^2\right) = \frac{1}{2}\left(\boldsymbol{T}_+\boldsymbol{T}_- + \boldsymbol{T}_-\boldsymbol{T}_+\right)$. Analogously, we make up $\frac{1}{4}\left(\boldsymbol{\lambda}_4^2+\boldsymbol{\lambda}_5^2\right)$ and $\frac{1}{4}\left(\boldsymbol{\lambda}_6^2+\boldsymbol{\lambda}_7^2\right)$. Using $\boldsymbol{\lambda}_8 = \frac{2}{\sqrt{3}}\left(\boldsymbol{U}_3+\boldsymbol{V}_3\right)$, (4.70), we obtain

$$\begin{aligned}\boldsymbol{C}_1 &= \frac{1}{2}\left(\boldsymbol{T}_+\boldsymbol{T}_- + \boldsymbol{T}_-\boldsymbol{T}_+ + \boldsymbol{V}_+\boldsymbol{V}_- + \boldsymbol{V}_-\boldsymbol{V}_+ + \boldsymbol{U}_+\boldsymbol{U}_- + \boldsymbol{U}_-\boldsymbol{U}_+\right)\\ &+ \frac{1}{4}\left(4\boldsymbol{T}_3^2 + \frac{4}{3}\left(\boldsymbol{V}_3+\boldsymbol{U}_3\right)^2\right).\end{aligned} \tag{4.83}$$

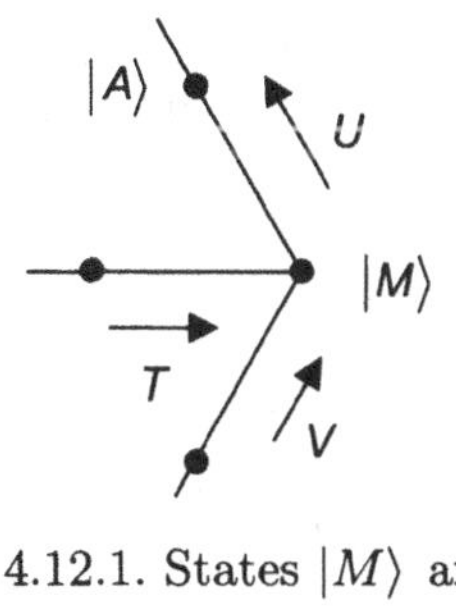

Figure 4.12.1. States $|M\rangle$ and $|A\rangle$.

As we mentioned in Section 4.11, all the states of a multiplet have the same eigenvalue of the Casimir operator. We choose arbitrarily the state $|M\rangle$ with the so-called maximal weight. According to (4.28) up to (4.30), we describe it by $|M\rangle = \left|\frac{p+q}{2}, \frac{p+q}{2}\right\rangle_T = \left|\frac{q}{2}, -\frac{q}{2}\right\rangle_U = \left|\frac{p}{2}, \frac{p}{2}\right\rangle_V$ and apply the operator $\boldsymbol{C}_1$ on it. If the operators $\boldsymbol{V}_+, \boldsymbol{T}_+$ and $\boldsymbol{U}_-$ act on $|M\rangle$ they lead out of the multiplet and yield zero. In order to make use of this property, we write, according to Table 4.2.1:

$$\begin{aligned} \boldsymbol{T}_+\boldsymbol{T}_- &= \boldsymbol{T}_-\boldsymbol{T}_+ + 2\boldsymbol{T}_3 \\ \boldsymbol{V}_+\boldsymbol{V}_- &= \boldsymbol{V}_-\boldsymbol{V}_+ + 2\boldsymbol{V}_3 \\ \boldsymbol{U}_-\boldsymbol{U}_+ &= \boldsymbol{U}_+\boldsymbol{U}_- - 2\boldsymbol{U}_3\,, \end{aligned} \tag{4.84}$$

which we insert in $\boldsymbol{C}_1$. We obtain

$$\boldsymbol{C}_1 = \boldsymbol{T}_-\boldsymbol{T}_+ + \boldsymbol{V}_-\boldsymbol{V}_+ + \boldsymbol{U}_+\boldsymbol{U}_- + \boldsymbol{T}_3 + \boldsymbol{V}_3 - \boldsymbol{U}_3 + \boldsymbol{T}_3^2 + \frac{1}{3}(\boldsymbol{V}_3 + \boldsymbol{U}_3)^2\,. \tag{4.85}$$

If $\boldsymbol{C}_1$ acts on $|M\rangle$, the first three expressions on the right hand side do not contribute to the result and we get

$$\begin{aligned} C_1|M\rangle &= \left(\frac{p+q}{2} + \frac{p}{2} + \frac{q}{2} + \left(\frac{p+q}{2}\right)^2 + \frac{1}{3}\left(\frac{p-q}{2}\right)^2\right)|M\rangle \\ &= \left(\frac{p^2+pq+q^2}{3} + p + q\right)|M\rangle\,. \end{aligned} \tag{4.86}$$

That is, the eigenvalue amounts to

$$C_1 = \frac{p^2+pq+q^2}{3} + p + q \tag{4.87}$$

up to an arbitrary factor, which depends on the definition of $\boldsymbol{C}_1$.

As an **exercise** and as a check, we calculate the Casimir eigenvalue by means of state $|A\rangle$ (see Figure 4.12.1). According to (4.31) up to (4.33) we write

$$|A\rangle = \left|\frac{p+q-1}{2}, \frac{p+q-1}{2}\right\rangle_T = \left|\frac{q}{2}, -\frac{q}{2}+1\right\rangle_U = \left|\frac{p+1}{2}, \frac{p+1}{2}\right\rangle_V\,. \tag{4.88}$$

By means of (3.65/66) we formulate

$$\begin{aligned} \boldsymbol{U}_+\boldsymbol{U}_-|A\rangle &= \boldsymbol{U}_+\sqrt{\frac{q}{2}\cdot\frac{q+2}{2} - \frac{q-2}{2}\cdot\frac{q}{2}}\,|M\rangle \\ &= \sqrt{q}\sqrt{\frac{q}{2}\cdot\frac{q+2}{2} - \frac{q-2}{2}\cdot\frac{q}{2}}\,|A\rangle = q\,|A\rangle \end{aligned} \tag{4.89}$$

and obtain

$$
\begin{aligned}
&C_1\,|A\rangle = \\
&\left(\boldsymbol{T}_-\boldsymbol{T}_+ + \boldsymbol{V}_-\boldsymbol{V}_+ + \boldsymbol{U}_+\boldsymbol{U}_- + \boldsymbol{T}_3 + \boldsymbol{V}_3 - \boldsymbol{U}_3 + \boldsymbol{T}_3^2 + \tfrac{1}{3}\left(\boldsymbol{V}_3 + \boldsymbol{U}_3\right)^2\right)|A\rangle = \\
&\left(0 + 0 + q + \tfrac{p+q-1}{2} + \tfrac{p+1}{2} + \tfrac{q-2}{2} + \left(\tfrac{p+q-1}{2}\right)^2 + \tfrac{1}{3}\left(\tfrac{p+1-q+2}{2}\right)^2\right)|A\rangle = \\
&\left(q + p + q - 1 + \tfrac{p^2+q^2+pq}{3} - q + 1\right)|A\rangle = \left(\tfrac{p^2+pq+q^2}{3} + p + q\right)|A\rangle\,,
\end{aligned}
\tag{4.90}
$$

which we expect because $|A\rangle$ belongs to the same multiplet as $|M\rangle$.

4.13 Direct products of $su(3)$-multiplets

In Sections 3.4/4.10 we have dealt with $su(2)/su(3)$ representations affiliated to $su(2)/su(3)$-multiplets. The direct product of irreducible $su(2)$-representations have been investigated in Section 3.5. In Section 3.6 their transformation in the block diagonal form, i.e., the decomposition in a direct sum of irreducible representations has been performed. By this transformation, the corresponding set of pairs of functions has been brought in a direct sum of multiplets of $su(2)$.

This procedure can be applied to $su(3)$-representations and -multiplets, which requires $su(3)$ Clebsch–Gordan coefficients. However, we state without proof that the decomposition of the $su(3)$ function set of a direct product of multiplets can be performed very similarly to the graphical method for $su(2)$ explained in Section 3.7.

Again, we depict the function set of a direct product in such a way that we draw the first multiplet $M(p,q)$ according to Figure 4.13.1 and set the second multiplet $M'(p',q')$ repeatedly so that its centre appears in every point of $M(p,q)$, which we delete afterwards.

For **example**, the tensor product $M_3\,(1,0) \otimes M_3'\,(1,0)$ (see Section 4.7) is made up like this:

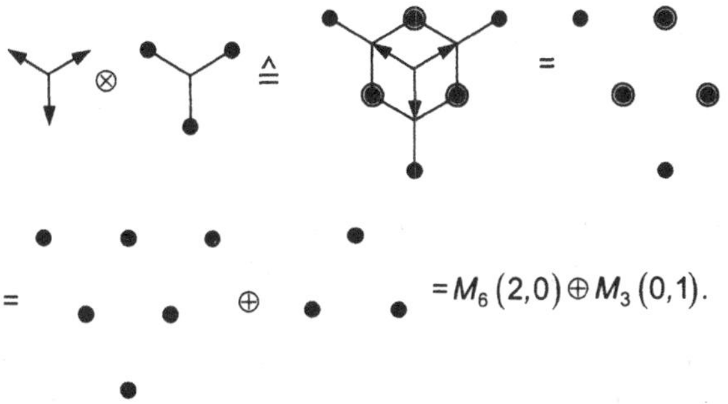

Figure 4.13.1. Tensor product of $M_3\,(1,0)$ and $M_3'(1,0)$.

In the second diagram each double occupancy is outlined by a point and a circle. The third diagram is decomposed by splitting off first the larger multiplet.

In Figures 4.13.2 and 4.13.3 **two other examples** follow:

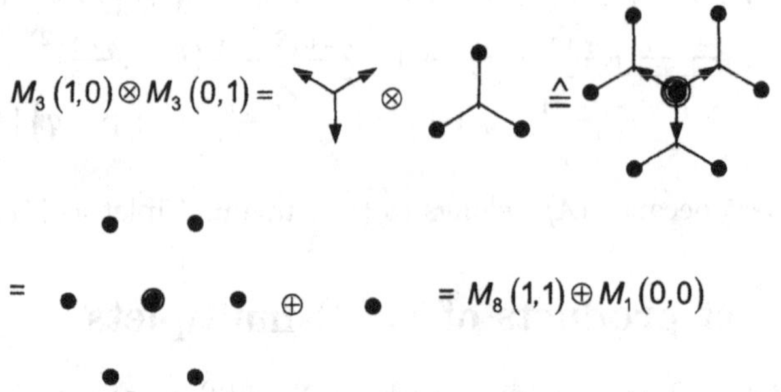

Figure 4.13.2. $M_3\,(1,0) \otimes M_3\,(0,1)$.

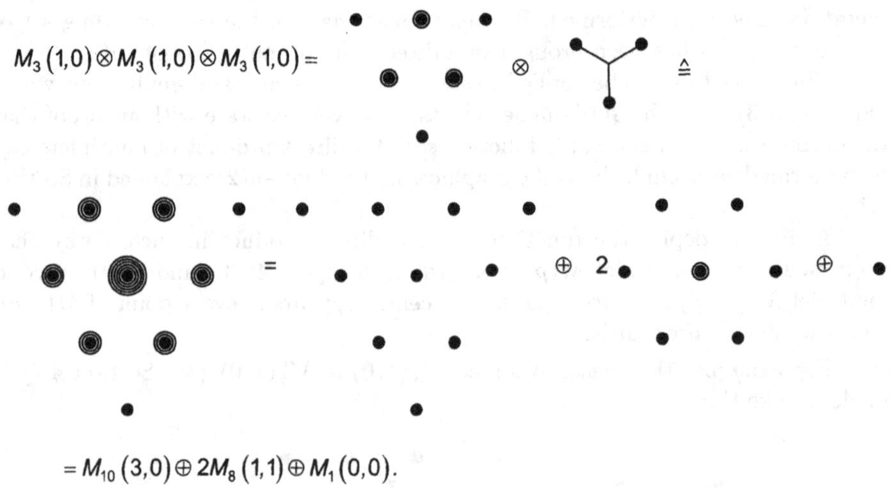

Figure 4.13.3. $M_3\,(1,0) \otimes M_3\,(1,0) \otimes M_3\,(1,0)$.

Here the diagram for $M_3\,(1,0) \otimes M_3\,(1,0)$ has been taken from the first example (Figure 4.13.1). The resulting direct sums in Figures 4.13.1 up to 4.13.3 are named *Clebsch–Gordan series.*

The states of one multiplet of a Clebsch–Gordan series differ in the coordinate Y from the states of another multiplet by $0, \pm 1, \pm 2, \ldots$. Therefore we state that *all multiplets of a Clebsch–Gordan series have the same triality* (see (4.67)).

In Table 4.13.1 the decomposition of the tensor products of the smallest multiplets is given.

Table 4.13.1. Direct products of $su(3)$-multiplets.

$$
\begin{aligned}
M_3(1,0)\otimes M_3(1,0) &\triangleq M_6(2,0)\oplus M_3(0,1)\\
M_3(1,0)\otimes M_3(0,1) &\triangleq M_8(1,1)\oplus M_1(0,0)\\
M_8(1,1)\otimes M_3(1,0) &\triangleq M_{15}(2,1)\oplus M_6(0,2)\oplus M_3(1,0)\\
M_8(1,1)\otimes M_8(1,1) &\triangleq M_{27}(2,2)\oplus M_{10}(3,0)\oplus M_{10}(0,3)\oplus 2M_8(1,1)\\
&\quad\oplus M_1(0,0)\\
M_6(2,0)\otimes M_3(1,0) &\triangleq M_{10}(3,0)\oplus M_8(1,1)\\
M_6(2,0)\otimes M_3(0,1) &\triangleq M_{15}(2,1)\oplus M_3(1,0)\\
M_6(2,0)\otimes M_8(1,1) &\triangleq M_{24}(3,1)\oplus M_{15}(1,2)\oplus M_6(2,0)\oplus M_3(1,0)\\
M_6(2,0)\otimes M_6(2,0) &\triangleq M_{15}(4,0)\oplus M_{15}(2,1)\oplus M_6(0,2)\\
M_6(2,0)\otimes M_6(0,2) &\triangleq M_{27}(2,2)\oplus M_8(1,1)\oplus M_1(0,0)\\
M_{15}(2,1)\otimes M_3(1,0) &\triangleq M_{24}(3,1)\oplus M_{15}(1,2)\oplus M_6(2,0)\\
M_{15}(2,1)\otimes M_3(0,1) &\triangleq M_{27}(2,2)\oplus M_{10}(3,0)\oplus M_8(1,1)\\
M_{15}(2,1)\otimes M_8(1,1) &\triangleq M_{42}(3,2)\oplus M_{24}(1,3)\oplus M_{15}(4,0)\oplus 2M_{15}(2,1)\\
&\quad\oplus M_6(0,2)\oplus M_3(1,0)\\
M_{15}(2,1)\otimes M_6(2,0) &\triangleq M_{35}(4,1)\oplus M_{27}(2,2)\oplus M_{10}(3,0)\oplus M_{10}(0,3)\\
&\quad\oplus M_8(1,1)\\
M_{15}(2,1)\otimes M_6(0,2) &\triangleq M_{42}(2,3)\oplus M_{24}(3,1)\oplus M_{15}(1,2)\oplus M_6(2,0)\\
&\quad\oplus M_3(0,1)\\
M_{15}(2,1)\otimes M_{15}(2,1) &\triangleq M_{60}(4,2)\oplus M_{21}(5,0)\oplus M_{42}(2,3)\oplus 2M_{24}(3,1)\\
&\quad\oplus M_{15}(0,4)\oplus 2M_{15}(1,2)\oplus M_6(2,0)\oplus M_3(0,1)\\
M_{15}(2,1)\otimes M_{15}(1,2) &\triangleq M_{64}(3,3)\oplus M_{35}(4,1)\oplus M_{35}(1,4)\oplus 2M_{27}(2,2)\\
&\quad\oplus 2M_{10}(0,3)\oplus 2M_8(1,1)\oplus M_1(0,0)\,.
\end{aligned}
$$

The subscript number is the dimension of the multiplet. The product of the dimensions of the factor multiplets must equal the sum of the dimensions of the decomposition.

4.14 Decomposition of direct products of multiplets by means of Young diagrams

In Section 4.6 we acquainted ourselves with the technique of Young diagrams determining the dimension of multiplets. We now apply it to direct products of multiplets in order to decompose them. Because the graphical method of Section 4.13 gets more and more laborious with growing values of p and q, the efficient method of Young is welcome.

We present the procedure of Young without proof and formulate it not only for two ordering numbers of p and q, but generally for $p_1, p_2, \ldots, p_{N-1}$ (the number $l = N-1$ of independent parameters is reviewed in Section 4.11). We work with an example containing two multiplets as factors, which have, however, only two parameter i.e. $N = 3$. We draw the factor multiplets in the standard arrangement (see Section 4.6) like in Figure 4.14.1.

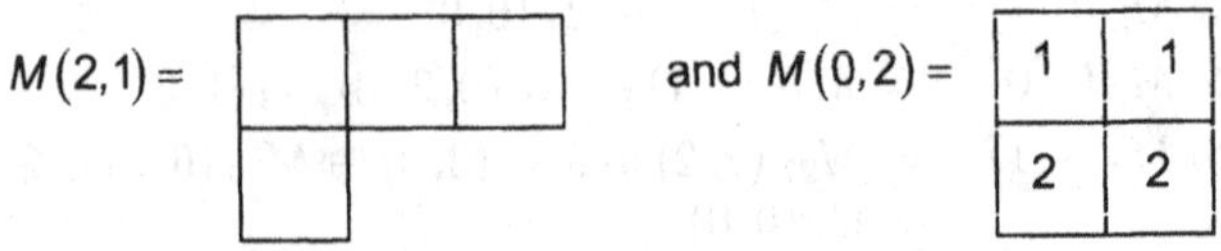

Figure 4.14.1. Standard arrangements.

In the second diagram we marked the boxes with the number of the corresponding row. We now add the boxes of the second diagram to the first one in all possible ways following these rules:

(1) No row may be longer than the one above.
(2) In a $su(N)$-multiplet no column may contain more than N boxes.
(3) Within a row, the numbers in the boxes must not decrease to the right.
(4) Within a column, the numbers must increase downwards.

In our example we begin with configurations which contain the number 1 in the first row like in Figure 4.14.2.

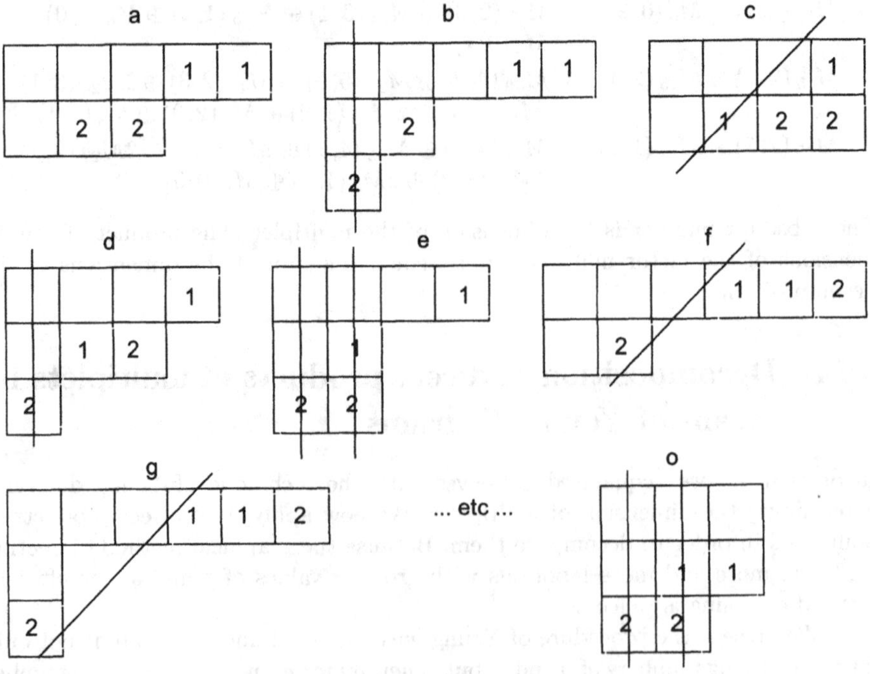

Figure 4.14.2. Configurations.

There are the following additional restrictions:

(5) In the resulting Young diagrams one goes through all boxes from the right to the left and from the top to the bottom. At each point of the path the number of boxes encountered with the number i must be less or equal to the number of boxes with $i-1$ (Figure 4.14.3).

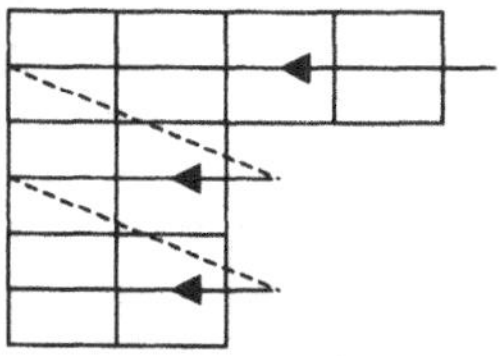

Figure 4.14.3. Path in the diagram.

Due to (5), in our example only the diagrams a,b,d,e and o are permitted.

(6) Columns with N boxes have to be deleted (if the diagram consists only of such columns there is the multiplet $M(0,0,\ldots,0)$).

In accordance with (6), the diagrams b,d,e and o have to be reduced. That is, we have the following result:

$$M_{15}\,(2,1)\otimes M_6\,(0,2) \triangleq M_{42}\,(2,3)\oplus M_{24}\,(3,1)\oplus M_{15}\,(1,2)\oplus M_6\,(2,0)\oplus M_3\,(0,1) \tag{4.91}$$

in accordance with Table 4.13.1. Another example reads:

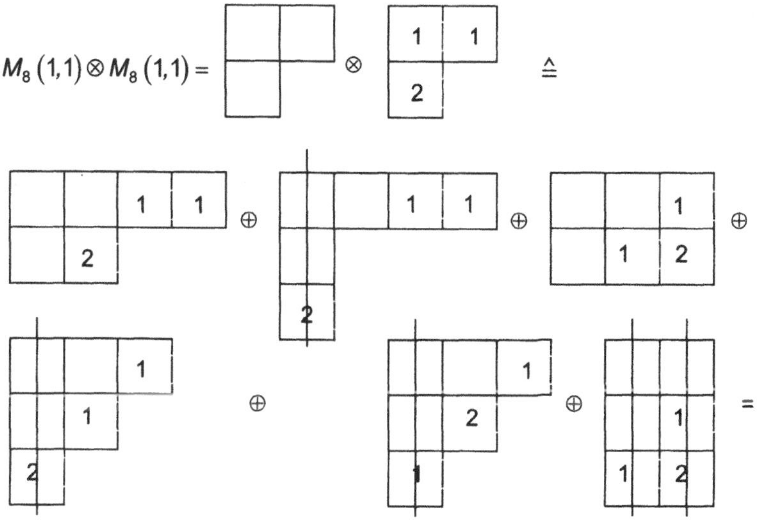

$M_{27}\,(2,2)\oplus M_{10}\,(3,0)\oplus M_{10}\,(0,3)\oplus 2\cdot M_8\,(1,1)\oplus M_1\,(0,0).$

Figure 4.14.4. Direct product $M_8\,(1,1)\otimes M_8\,(1,1)$.

Of course, the technique of Young applies also to $su(2)$-multiplets. In connection with expression (3.60), we have denoted them by $[j]$ or $M_{2j+1}(2j)$. The second symbol shows that its dimension is $2j + 1$ and the number of states minus one is $2j$ (see (4.27)). In the following example we will start with the $[j]$-notation and finally we will come back to it (Figure 4.14.5).

$$M_6(5) \oplus M_4(3) \oplus M_2(1) \equiv \left[\frac{5}{2}\right] \oplus \left[\frac{3}{2}\right] \oplus \left[\frac{1}{2}\right]$$

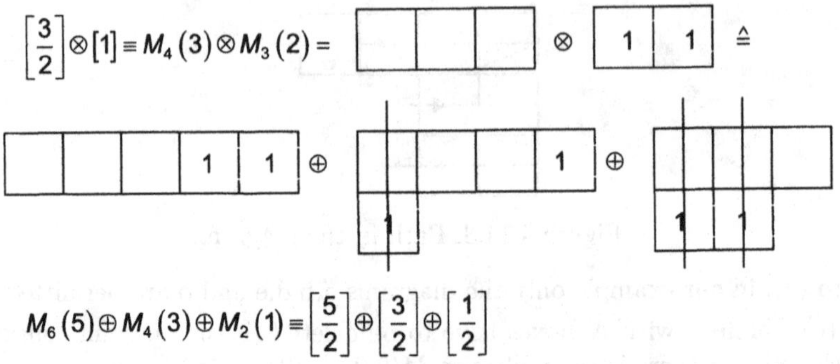

Figure 4.14.5. Direct product $M_4(3) \otimes M_3(2)$.

The result is in accordance with (3.121).

Chapter 5

The Lie algebra $su(4)$

Analogously to the previous chapter, the basis elements, generators and structure constants of $su(4)$ are determined and discussed. There are three additional subalgebras in $su(4)$, the step- and "third" operators of which determine the "three dimensional" structure of the multiplets. Their decomposition in "layers" and their dimension are dealt with.

In analogy with the "hypercharge", the so-called "charm" is introduced. Direct products of multiplets and their decomposition in Clebsch–Gordan series are performed mainly by means of Young diagrams.

The operators constituting the Cartan–Weyl basis of $su(4)$ are given. The canonical relations, which comprise a set of rules for the commutators of these operators, are presented numerically.

5.1 The generators of the $su(4)$-algebra, subalgebras

The anti-Hermitian 4×4-matrices $\underline{\underline{e}}_i$ with vanishing traces constitute the basis of $su(4)$. In analogy to Sections 2.2 and 4.1, Hermitian 4×4-matrices $\underline{\underline{\lambda}}_i$ or generators are split off from $\underline{\underline{e}}_i$ this way: $\underline{\underline{e}}_i = -\frac{1}{2}\mathrm{i}\underline{\underline{\lambda}}_i$.

Due to (2.4/2.5), the seven λ-matrices from (4.1) can be used attaching a column of zeros on the right in the matrices and a row of zeros at the bottom. We proceed in the same way with the diagonal matrix of (4.2) and obtain

$$
\begin{aligned}
&\underline{\underline{\lambda}}_1 = \begin{pmatrix} 0 & 1 & 0 & 0 \\ 1 & 0 & 0 & 0 \\ 0 & 0 & 0 & 0 \\ 0 & 0 & 0 & 0 \end{pmatrix}, \quad
\underline{\underline{\lambda}}_2 = \begin{pmatrix} 0 & -i & 0 & 0 \\ i & 0 & 0 & 0 \\ 0 & 0 & 0 & 0 \\ 0 & 0 & 0 & 0 \end{pmatrix}, \\
&\underline{\underline{\lambda}}_3 = \begin{pmatrix} 1 & 0 & 0 & 0 \\ 0 & -1 & 0 & 0 \\ 0 & 0 & 0 & 0 \\ 0 & 0 & 0 & 0 \end{pmatrix}, \\
&\underline{\underline{\lambda}}_4 = \begin{pmatrix} 0 & 0 & 1 & 0 \\ 0 & 0 & 0 & 0 \\ 1 & 0 & 0 & 0 \\ 0 & 0 & 0 & 0 \end{pmatrix}, \quad
\underline{\underline{\lambda}}_5 = \begin{pmatrix} 0 & 0 & -i & 0 \\ 0 & 0 & 0 & 0 \\ i & 0 & 0 & 0 \\ 0 & 0 & 0 & 0 \end{pmatrix}, \\
&\underline{\underline{\lambda}}_6 = \begin{pmatrix} 0 & 0 & 0 & 0 \\ 0 & 0 & 1 & 0 \\ 0 & 1 & 0 & 0 \\ 0 & 0 & 0 & 0 \end{pmatrix}, \quad
\underline{\underline{\lambda}}_7 = \begin{pmatrix} 0 & 0 & 0 & 0 \\ 0 & 0 & -i & 0 \\ 0 & i & 0 & 0 \\ 0 & 0 & 0 & 0 \end{pmatrix}, \\
&\underline{\underline{\lambda}}_8 = \frac{1}{\sqrt{3}} \begin{pmatrix} 1 & 0 & 0 & 0 \\ 0 & 1 & 0 & 0 \\ 0 & 0 & -2 & 0 \\ 0 & 0 & 0 & 0 \end{pmatrix}.
\end{aligned}
\tag{5.1}
$$

According to (2.3) we complete the set with

$$
\begin{aligned}
&\underline{\underline{\lambda}}_9 = \begin{pmatrix} 0 & 0 & 0 & 1 \\ 0 & 0 & 0 & 0 \\ 0 & 0 & 0 & 0 \\ 1 & 0 & 0 & 0 \end{pmatrix}, \quad
\underline{\underline{\lambda}}_{10} = \begin{pmatrix} 0 & 0 & 0 & -i \\ 0 & 0 & 0 & 0 \\ 0 & 0 & 0 & 0 \\ i & 0 & 0 & 0 \end{pmatrix}, \\
&\underline{\underline{\lambda}}_{11} = \begin{pmatrix} 0 & 0 & 0 & 0 \\ 0 & 0 & 0 & 1 \\ 0 & 0 & 0 & 0 \\ 0 & 1 & 0 & 0 \end{pmatrix}, \quad
\underline{\underline{\lambda}}_{12} = \begin{pmatrix} 0 & 0 & 0 & 0 \\ 0 & 0 & 0 & -i \\ 0 & 0 & 0 & 0 \\ 0 & i & 0 & 0 \end{pmatrix}, \\
&\underline{\underline{\lambda}}_{13} = \begin{pmatrix} 0 & 0 & 0 & 0 \\ 0 & 0 & 0 & 0 \\ 0 & 0 & 0 & 1 \\ 0 & 0 & 1 & 0 \end{pmatrix}, \quad
\underline{\underline{\lambda}}_{14} = \begin{pmatrix} 0 & 0 & 0 & 0 \\ 0 & 0 & 0 & 0 \\ 0 & 0 & 0 & -i \\ 0 & 0 & i & 0 \end{pmatrix}.
\end{aligned}
\tag{5.2}
$$

In order to meet the trace condition (2.27) (see (4.2)), we choose the remaining matrix like this:

$$
\underline{\underline{\lambda}}_{15} = \frac{1}{\sqrt{6}} \begin{pmatrix} 1 & 0 & 0 & 0 \\ 0 & 1 & 0 & 0 \\ 0 & 0 & 1 & 0 \\ 0 & 0 & 0 & -3 \end{pmatrix}. \tag{5.3}
$$

The commutator relation (2.26) applies $\left[\underline{\underline{\lambda}}_i, \underline{\underline{\lambda}}_k\right] = \sum\limits_l C_{ikl} 2\mathrm{i}\lambda_l$. The **structure constants** connecting the generators $\underline{\underline{\lambda}}_1$ up to $\underline{\underline{\lambda}}_8$ are identical to the values in Table 4.1.1. For the residual generators the following structure constants result:

Table 5.1.1. Additional non-vanishing structure constants of $su(4)$.

i	k	l	C_{ikl}	i	k	l	C_{ikl}
1	9	12	1/2	6	11	14	1/2
1	10	11	−1/2	6	12	13	−1/2
2	9	11	1/2	7	11	13	1/2
2	10	12	1/2	7	12	14	1/2
3	9	10	1/2	8	9	10	$\sqrt{1/12}$
3	11	12	−1/2	8	11	12	$\sqrt{1/12}$
4	9	14	1/2	8	13	14	$-\sqrt{1/3}$
4	10	13	−1/2	9	10	15	$\sqrt{2/3}$
5	9	13	1/2	11	12	15	$\sqrt{2/3}$
5	10	14	1/2	13	14	15	$\sqrt{2/3}$.

From Tables 4.1.1 and 5.1.1 we take

$$\left[\underline{\underline{\lambda}}_3, \underline{\underline{\lambda}}_8\right] = \left[\underline{\underline{\lambda}}_3, \underline{\underline{\lambda}}_{15}\right] = \left[\underline{\underline{\lambda}}_8, \underline{\underline{\lambda}}_{15}\right] = 0, \tag{5.4}$$

that is, the generators $\underline{\underline{\lambda}}_3$, $\underline{\underline{\lambda}}_8$ and $\underline{\underline{\lambda}}_{15}$ (and not more) commute mutually. Therefore, the Lie algebra $su(4)$ has rank 3 (see at the end of Section 4.1 or the second paragraph of 4.11). Due to the strict correspondence between the generators $\underline{\underline{\lambda}}_1$ up to $\underline{\underline{\lambda}}_7$ of the algebra $su(3)$ and the same generators of $su(4)$, $su(3)$ is a subalgebra of $su(4)$. Generally, the algebras $su(N')$ with $N' < N$ are subalgebras of $su(N)$.

From now on, we deal with the operators $\boldsymbol{\lambda}_1, \boldsymbol{\lambda}_2, \ldots, \boldsymbol{\lambda}_{15}$ constituting an algebra which is isomorphic to the matrix representation $\left\{\underline{\underline{\lambda}}_1, \underline{\underline{\lambda}}_2, \ldots, \underline{\underline{\lambda}}_{15}\right\}$. Their commutators reveal also the structure constants of Tables 4.1.1 and 5.1.1. In $su(4)$ the $\boldsymbol{T}$-, $\boldsymbol{U}$- and $\boldsymbol{V}$-operators are defined identically to (4.9) up to (4.11).

In addition, we define the following **operators**:

$$\begin{aligned}
\boldsymbol{W}_\pm &= \tfrac{1}{2}\left(\boldsymbol{\lambda}_9 \pm \mathrm{i}\boldsymbol{\lambda}_{10}\right), & \boldsymbol{W}_3 &= \tfrac{1}{2}\left(\tfrac{1}{2}\boldsymbol{\lambda}_3 + \tfrac{1}{2\sqrt{3}}\boldsymbol{\lambda}_8 + \sqrt{\tfrac{2}{3}}\boldsymbol{\lambda}_{15}\right), \\
\boldsymbol{X}_\pm &= \tfrac{1}{2}\left(\boldsymbol{\lambda}_{11} \pm \mathrm{i}\boldsymbol{\lambda}_{12}\right), & \boldsymbol{X}_3 &= \tfrac{1}{2}\left(-\tfrac{1}{2}\boldsymbol{\lambda}_3 + \tfrac{1}{2\sqrt{3}}\boldsymbol{\lambda}_8 + \sqrt{\tfrac{2}{3}}\boldsymbol{\lambda}_{15}\right), \\
\boldsymbol{Z}_\pm &= \tfrac{1}{2}\left(\boldsymbol{\lambda}_{13} \pm \mathrm{i}\boldsymbol{\lambda}_{14}\right), & \boldsymbol{Z}_3 &= \tfrac{1}{2}\left(\qquad -\tfrac{1}{\sqrt{3}}\boldsymbol{\lambda}_8 + \sqrt{\tfrac{2}{3}}\boldsymbol{\lambda}_{15}\right)
\end{aligned} \tag{5.5}$$

with the following linear dependencies

$$\boldsymbol{T}_3 + \boldsymbol{X}_3 = \boldsymbol{W}_3, \quad \boldsymbol{U}_3 = \boldsymbol{X}_3 - \boldsymbol{Z}_3. \tag{5.6}$$

By means of the structure constants, the following relations can be shown:

$$\begin{aligned}
[\boldsymbol{W}_3, \boldsymbol{W}_\pm] &= \pm \boldsymbol{W}_\pm, & [\boldsymbol{W}_+, \boldsymbol{W}_-] &= 2\boldsymbol{W}_3, \\
[\boldsymbol{X}_3, \boldsymbol{X}_\pm] &= \pm \boldsymbol{X}_\pm, & [\boldsymbol{X}_+, \boldsymbol{X}_-] &= 2\boldsymbol{X}_3, \\
[\boldsymbol{Z}_3, \boldsymbol{Z}_\pm] &= \pm \boldsymbol{Z}_\pm, & [\boldsymbol{Z}_+, \boldsymbol{Z}_-] &= 2\boldsymbol{Z}_3,
\end{aligned} \tag{5.7}$$

in analogy to (4.10) and (4.12).

As explained in the context of eq. (4.13), $su(2)$-multiplets are affiliated to the T-, U-, V-, W-, X- and Z-algebras. The commutators of the $\boldsymbol{T}$-, $\boldsymbol{U}$- and $\boldsymbol{V}$-operators are given in Table 4.2.1. In Table 5.1.2 we put together the commutators of the $\boldsymbol{W}$-, $\boldsymbol{X}$- and $\boldsymbol{Z}$- operators with all operators of this kind.

Table 5.1.2. Commutator relations of $\boldsymbol{W}$-, $\boldsymbol{X}$- and $\boldsymbol{Z}$-operators with $\boldsymbol{T}$-, $\boldsymbol{U}$-, $\boldsymbol{V}$-, $\boldsymbol{W}$-, $\boldsymbol{X}$- and $\boldsymbol{Z}$-operators. The commutator of an operator in the top line with an operator in the column at the right hand side results in the operator given in the corresponding field.

$\boldsymbol{W}_+$	$\boldsymbol{W}_-$	$\boldsymbol{W}_3$	$\boldsymbol{X}_+$	$\boldsymbol{X}_-$	$\boldsymbol{X}_3$	$\boldsymbol{Z}_+$	$\boldsymbol{Z}_-$	$\boldsymbol{Z}_3$	
0	$\boldsymbol{X}_-$	$\frac{1}{2}\boldsymbol{T}_+$	$-\boldsymbol{W}_+$	0	$-\frac{1}{2}\boldsymbol{T}_+$	0	0	0	$\boldsymbol{T}_+$
$-\boldsymbol{X}_+$	0	$-\frac{1}{2}\boldsymbol{T}_-$	0	$\boldsymbol{W}_-$	$\frac{1}{2}\boldsymbol{T}_-$	0	0	0	$\boldsymbol{T}_-$
$-\frac{1}{2}\boldsymbol{W}_+$	$\frac{1}{2}\boldsymbol{W}_-$	0	$\frac{1}{2}\boldsymbol{X}_+$	$-\frac{1}{2}\boldsymbol{X}_-$	0	0	0	0	$\boldsymbol{T}_3$
0	0	0	0	$\boldsymbol{Z}_-$	$\frac{1}{2}\boldsymbol{U}_+$	$-\boldsymbol{X}_+$	0	$-\frac{1}{2}\boldsymbol{U}_+$	$\boldsymbol{U}_+$
0	0	0	$-\boldsymbol{Z}_+$	0	$-\frac{1}{2}\boldsymbol{U}_-$	0	$\boldsymbol{X}_-$	$\frac{1}{2}\boldsymbol{U}_-$	$\boldsymbol{U}_-$
0	0	0	$-\frac{1}{2}\boldsymbol{X}_+$	$\frac{1}{2}\boldsymbol{X}_-$	0	$\frac{1}{2}\boldsymbol{Z}_+$	$-\frac{1}{2}\boldsymbol{Z}_-$	0	$\boldsymbol{U}_3$
0	$\boldsymbol{Z}_-$	$\frac{1}{2}\boldsymbol{V}_+$	0	0	0	$-\boldsymbol{W}_+$	0	$-\frac{1}{2}\boldsymbol{V}_+$	$\boldsymbol{V}_+$
$-\boldsymbol{Z}_+$	0	$-\frac{1}{2}\boldsymbol{V}_-$	0	0	0	0	$\boldsymbol{W}_-$	$\frac{1}{2}\boldsymbol{V}_-$	$\boldsymbol{V}_-$
$-\frac{1}{2}\boldsymbol{W}_+$	$\frac{1}{2}\boldsymbol{W}_-$	0	0	0	0	$\frac{1}{2}\boldsymbol{Z}_+$	$-\frac{1}{2}\boldsymbol{Z}_-$	0	$\boldsymbol{V}_3$
	$-2\boldsymbol{W}_3$	$\boldsymbol{W}_+$	0	$-\boldsymbol{T}_+$	$\frac{1}{2}\boldsymbol{W}_+$	0	$-\boldsymbol{V}_+$	$\frac{1}{2}\boldsymbol{W}_+$	$\boldsymbol{W}_+$
		$-\boldsymbol{W}_-$	$\boldsymbol{T}_-$	0	$-\frac{1}{2}\boldsymbol{W}_-$	$\boldsymbol{V}_-$	0	$-\frac{1}{2}\boldsymbol{W}_-$	$\boldsymbol{W}_-$
			$-\frac{1}{2}\boldsymbol{X}_+$	$\frac{1}{2}\boldsymbol{X}_-$	0	$-\frac{1}{2}\boldsymbol{Z}_+$	$\frac{1}{2}\boldsymbol{Z}_-$	0	$\boldsymbol{W}_3$
				$-2\boldsymbol{X}_3$	$\boldsymbol{X}_+$	0	$-\boldsymbol{U}_+$	$\frac{1}{2}\boldsymbol{X}_+$	$\boldsymbol{X}_+$
					$-\boldsymbol{X}_-$	$\boldsymbol{U}_-$	0	$-\frac{1}{2}\boldsymbol{X}_-$	$\boldsymbol{X}_-$
						$-\frac{1}{2}\boldsymbol{Z}_+$	$\frac{1}{2}\boldsymbol{Z}_-$	0	$\boldsymbol{X}_3$
							$-2\boldsymbol{Z}_3$	$\boldsymbol{Z}_+$	$\boldsymbol{Z}_+$
								$-\boldsymbol{Z}_-$	$\boldsymbol{Z}_-$

5.2 Step operators and states in $su(4)$

From Tables 4.2.1 and 5.1.2 we observe that the operators $\boldsymbol{T}_3$, $\boldsymbol{U}_3$, $\boldsymbol{V}_3$, $\boldsymbol{W}_3$, $\boldsymbol{X}_3$ and $\boldsymbol{Z}_3$ mutually commute. Therefore, every function of $su(4)$ which is an eigenfunction of one of the operators is simultaneously an eigenfunction of all of them and can be written this way:

$$|T_3\, U_3\, V_3\, W_3\, X_3\, Z_3\rangle\,. \tag{5.8}$$

Table 5.1.2 reveals the relations

$$\begin{aligned}[\boldsymbol{W}_3, \boldsymbol{T}_\pm] &= \pm\frac{1}{2}\boldsymbol{T}_\pm,\\ [\boldsymbol{X}_3, \boldsymbol{T}_\pm] &= \mp\frac{1}{2}\boldsymbol{T}_\pm,\\ [\boldsymbol{Z}_3, \boldsymbol{T}_\pm] &= 0.\end{aligned} \tag{5.9}$$

In analogy with (4.19) up to (4.23), we infer

$$\begin{aligned}&\boldsymbol{T}_\pm\,|T_3\, U_3\, V_3\, W_3\, X_3\, Z_3\rangle =\\ &\quad\sum_{X_3'\, Z_3'} N'(X_3', Z_3')\,\big|T_3\pm 1,\, U_3\mp\tfrac{1}{2},\, V_3\pm\tfrac{1}{2},\, W_3\pm\tfrac{1}{2},\, X_3'\, Z_3'\big\rangle,\\ &\boldsymbol{T}_\pm\,|T_3\, U_3\, V_3\, W_3\, X_3\, Z_3\rangle =\\ &\quad\sum_{W_3''\, Z_3''} N''(W_3'', Z_3'')\,\big|T_3\pm 1,\, U_3\mp\tfrac{1}{2},\, V_3\pm\tfrac{1}{2},\, W_3'',\, X_3\mp\tfrac{1}{2},\, Z_3''\big\rangle,\\ &\boldsymbol{T}_\pm\,|T_3\, U_3\, V_3\, W_3\, X_3\, Z_3\rangle =\\ &\quad\sum_{W_3'''\, X_3'''} N'''(W_3''', X_3''')\,\big|T_3\pm 1,\, U_3\mp\tfrac{1}{2},\, V_3\pm\tfrac{1}{2},\, W_3''',\, X_3''',\, Z_3\big\rangle,\end{aligned} \tag{5.10}$$

where we made use of (4.23). We sum up:

$$\boldsymbol{T}_\pm\,|T_3\, U_3\, V_3\, W_3\, X_3\, Z_3\rangle \to \left|T_3\pm 1,\, U_3\mp\frac{1}{2},\, V_3\pm\frac{1}{2},\, W_3\pm\frac{1}{2},\, X_3\mp\frac{1}{2},\, Z_3\right\rangle. \tag{5.11}$$

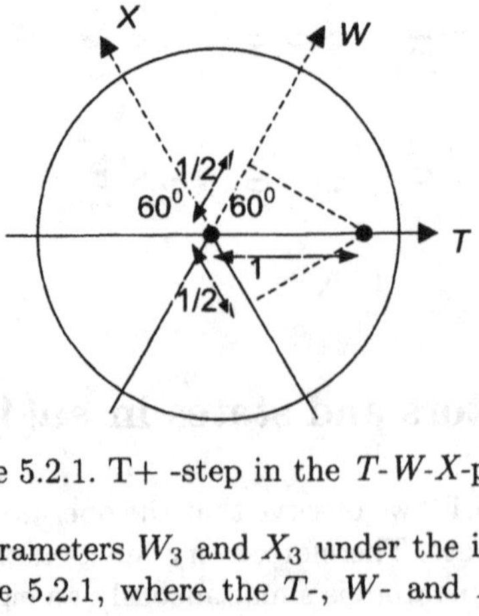

Figure 5.2.1. T+ -step in the T-W-X-plane.

The behaviour of the parameters W_3 and X_3 under the influence of the operator $\boldsymbol{T}_\pm$ is outlined in Figure 5.2.1, where the T-, W- and X-line constitute a new plane, which is indicated by a big circle. Because Z_3 is not changed in (5.11), the Z-line must be perpendicular to the T-line. Analogously to (5.11), one derives

$$\begin{aligned} \boldsymbol{U}_\pm \,|T_3\, U_3\, V_3\, W_3\, X_3\, Z_3\rangle &\to \left|T_3 \mp \tfrac{1}{2},\, U_3 \pm 1,\, V_3 \pm \tfrac{1}{2},\, W_3,\, X_3 \pm \tfrac{1}{2},\, Z_3 \mp \tfrac{1}{2}\right\rangle, \\ \boldsymbol{V}_\pm \,|T_3\, U_3\, V_3\, W_3\, X_3\, Z_3\rangle &\to \left|T_3 \pm \tfrac{1}{2},\, U_3 \pm \tfrac{1}{2},\, V_3 \pm 1,\, W_3 \pm \tfrac{1}{2},\, X_3,\, Z_3 \mp \tfrac{1}{2}\right\rangle. \end{aligned} \tag{5.12}$$

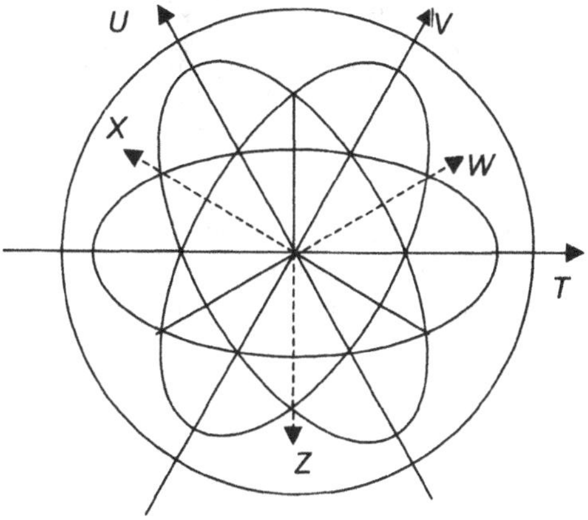

Figure 5.2.2. Lines and planes of the $su(4)$-multiplets.

That is, in order to represent the action of these operators geometrically we need an additional UXZ-plane and a VWZ-plane. The mentioned three planes are drawn as ellipses in Figure 5.2.2. The W-, X- and Z-axes point downward below the drawing plane, which is stepped up by the T-, U- and V-line and marked by the large circle. For example, the action of the operator $\boldsymbol{T}_+$, which is drawn in Figures 4.3.1 and 5.2.1, can also be perceived in Figure 5.2.2.

From (5.11) and (5.12) we expect that the line pairs (T, Z), (V, X) and (U, W) are orthogonal, which is realised in Figure 5.2.2.

Considerations analogous to the derivations of (5.11) and (5.12) lead to

$$\begin{aligned} \boldsymbol{W}_\pm \,|T_3\, U_3\, V_3\, W_3\, X_3\, Z_3\rangle &\to \left|T_3 \pm \tfrac{1}{2},\, U_3,\, V_3 \pm \tfrac{1}{2},\, W_3 \pm 1,\, X_3 \pm \tfrac{1}{2},\, Z_3 \pm \tfrac{1}{2}\right\rangle, \\ \boldsymbol{X}_\pm \,|T_3\, U_3\, V_3\, W_3\, X_3\, Z_3\rangle &\to \left|T_3 \mp \tfrac{1}{2},\, U_3 \pm \tfrac{1}{2},\, V_3,\, W_3 \pm \tfrac{1}{2},\, X_3 \pm 1,\, Z_3 \pm \tfrac{1}{2}\right\rangle, \\ \boldsymbol{Z}_\pm \,|T_3\, U_3\, V_3\, W_3\, X_3\, Z_3\rangle &\to \left|T_3,\, U_3 \mp \tfrac{1}{2},\, V_3 \mp \tfrac{1}{2},\, W_3 \pm \tfrac{1}{2},\, X_3 \pm \tfrac{1}{2},\, Z_3 \pm 1\right\rangle. \end{aligned} \tag{5.13}$$

The relations (5.13) can also be observed in Figure 5.2.2.

5.3 Multiplets of $su(4)$

Repeated applications of the operators $\boldsymbol{T}_\pm$, $\boldsymbol{U}_\pm$, $\boldsymbol{V}_\pm$, $\boldsymbol{W}_\pm$, $\boldsymbol{X}_\pm$ and $\boldsymbol{Z}_\pm$ generate a lattice of states in a tetrahedron like formation in analogy to the triangular formation in $su(3)$. Figure 5.3.1 displays the tetrahedrons formed by two layers of lattice points. The arrows point downward to the circles in the drawing plane.

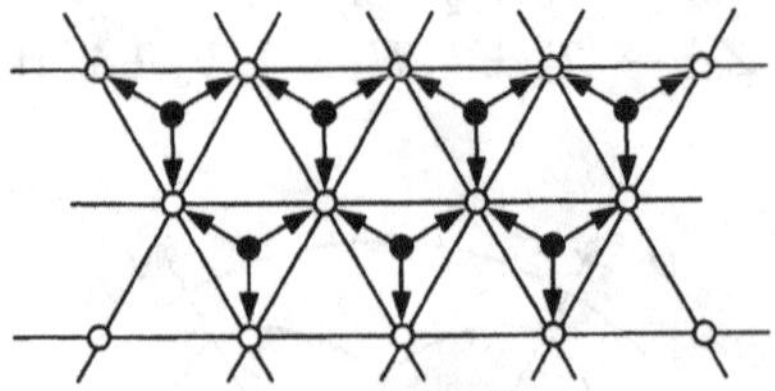

Figure 5.3.1. Two layers of the $su(4)$-lattice.

Similar considerations as in Section 4.4 show that a $su(4)$-multiplet is a finite and convex part of this lattice. Symmetry arguments as for $su(3)$ yield that a $su(4)$-multiplet $M_d(p_1, p_2, p_3)$ is constructed with the help of the following recipe:

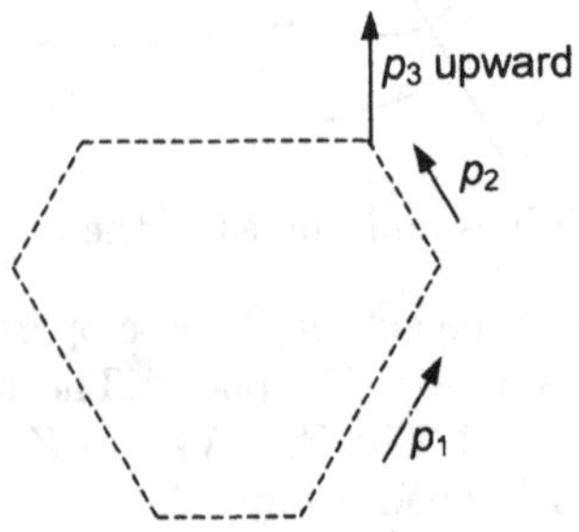

Figure 5.3.2. Main edges of the $su(4)$-multiplet $M_d(p_1, p_2, p_3)$.

a) The lowest layer (marked by circles) is a $su(3)$-multiplet with ordering numbers p_1 and p_2 (instead of p and q). At the end of the second line (with p_2) climb up p_3 units in the direction $-Z$ (see the raising arrow in Figure 5.3.2).

b) Starting from the point just reached, draw the smallest $su(3)$-multiplet in the "horizontal" plane with the same symmetry centre as the multiplet in the basal plane.

c) Add further $su(3)$-multiplets in higher planes in such a way that all lattice points are also members of $su(3)$-multiplets lying in the sloping planes TWX, UXZ or VWZ (see Figure 5.2.2).

d) The body of the $su(4)$-multiplet has to be convex.

e) Extending the recipe for "short paths" (see the end of Section 4.5) we say that if there are n "short paths" starting inward from a vertex the multiplicity is increased by $n - 1$. Symmetric paths are taken as one path, and a path leading over a doubly occupied boundary point is equivalent to two paths.

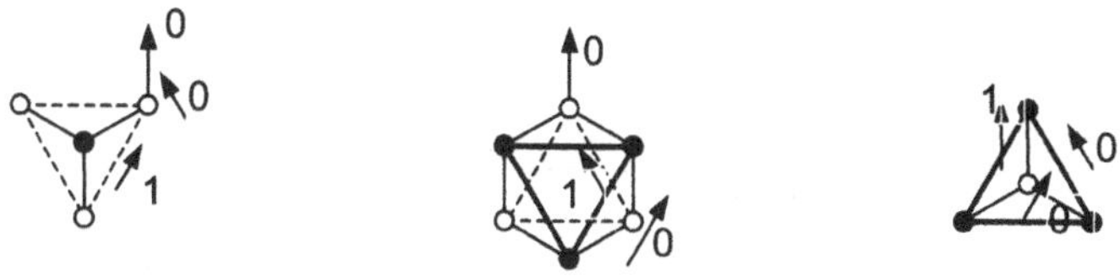

$M_4(1,0,0)$ $M_6(0,1,0)$ $M_4(0,0,1)$

Figure 5.3.3. $su(4)$-multiplets, $p_i = p_k = 0, p_l = 1$.

In Figure 5.3.3 the **smallest *su*(4)-multiplets** are shown. The $su(3)$-multiplets in the basal plane are drawn with dashed lines and circles. In M_4(0,0,1) it consists of a single state. The multiplet M_4(0,0,1) results from M_4(1,0,0) by a 180^0 rotation around the T-line. Generally, such a rotation transforms $M_d\ (p,q,r)$ in $M_d\ (r,q,p)$.

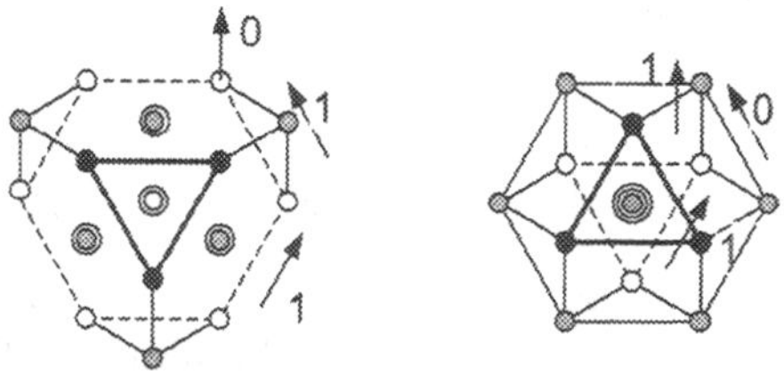

$M_{20}(1,1,0)$ $M_{15}(1,0,1)$

Figure 5.3.4a. $su(4)$-multiplets, $p_i = p_k = 1, p_l = 0$.

The representation of M_{20}(0,1,1) can be deduced from M_{20}(1,1,0) by the rotation mentioned above.

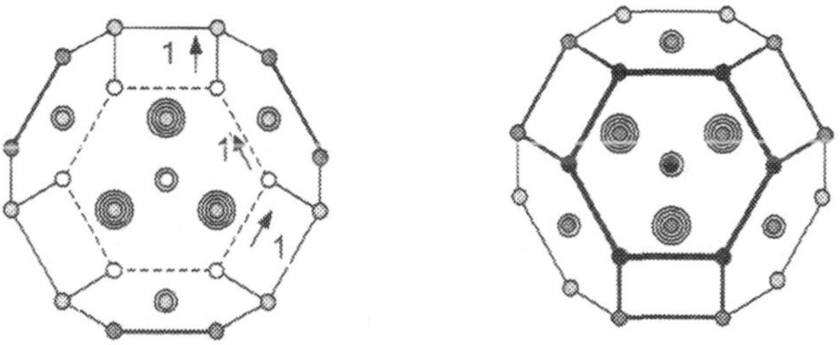

$M_{64}(1,1,1)$, lower part, $M_{64}(1,1,1)$, upper part.

Figure 5.3.4b.

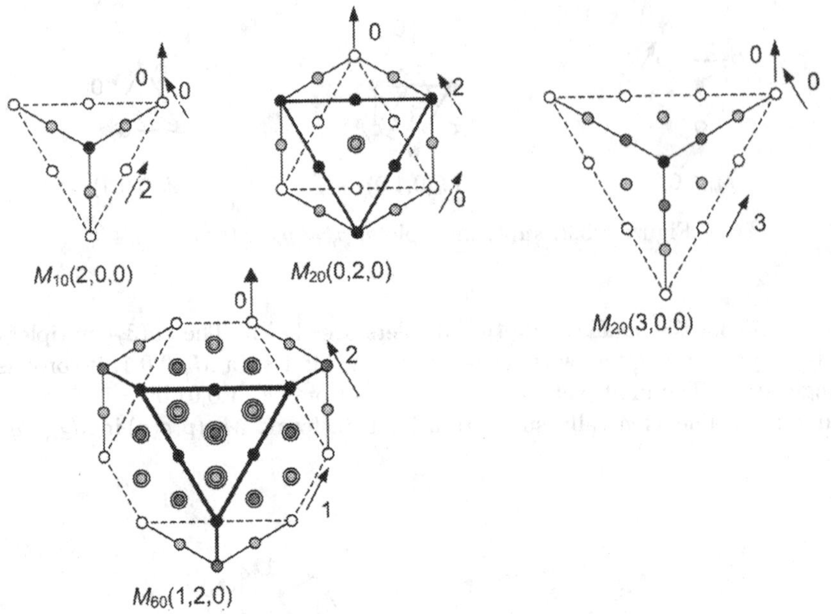

Figure 5.3.5. $su(4)$ multiplets, $p_i > 1$, $p_k = 1/0$, $p_3 = 0$.

In the outlines of $M_{20}(3,0,0)$ and $M_{60}(1,2,0)$ some state points are blocked out.

In order to clarify the structure of ***su*(4)-multiplets**, they are **decomposed in layers** parallel to the basal plane. The layers themselves are split up in $su(3)$-multiplets. In Figure 5.3.6 for $M_{60}(1,2,0)$ the first layer above the basal plane ($C = 1$) and the second ($C = 2$) are given.

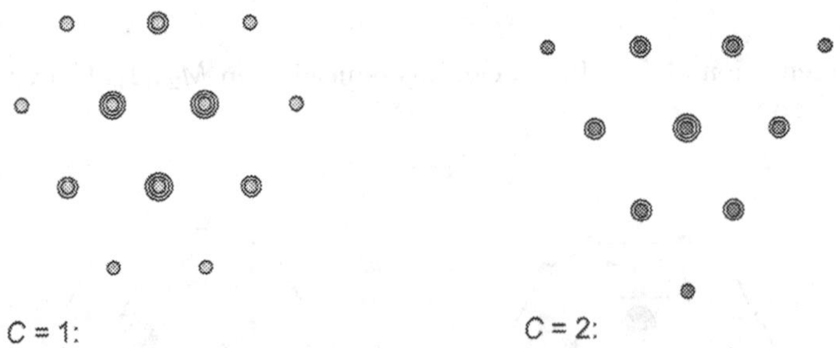

Figure 5.3.6. Intermediate layers of the multiplet $M_{60}(1,2,0)$.

In Table 5.3.1 the decompositions of several $su(4)$-multiplets in $su(3)$-multiplets is listed. Superscript numbers mark the numbers (above the basal plane) of the layers.

Table 5.3.1. Decomposition of $su(4)$-multiplets in $su(3)$-multiplets.

$$
\begin{aligned}
M_4\,(1,0,0) &= M_3^{(0)}\,(1,0) \oplus M_1^{(1)}\,(0,0)\,, \\
M_6\,(0,1,0) &= M_3^{(0)}\,(0,1) \oplus M_3^{(1)}\,(1,0)\,, \\
M_4\,(0,0,1) &= M_1^{(0)}\,(0,0) \oplus M_3^{(1)}\,(0,1)\,, \\
M_{20}\,(1,1,0) &= M_8^{(0)}\,(1,1) \oplus M_6^{(1)}\,(2,0) \oplus M_3^{(1)}\,(0,1) \oplus M_3^{(2)}\,(1,0)\,, \\
M_{15}\,(1,0,1) &= M_3^{0}\,(1,0) \;\oplus M_8^{(1)}\,(1,1) \oplus M_1^{(1)}\,(0,0) \oplus M_3^{(2)}\,(0,1)\,, \\
M_{64}\,(1,1,1) &= M_8^{(0)}\,(1,1) \oplus M_{15}^{(1)}\,(1,2) \oplus M_6^{(1)}\,(2,0) \oplus M_3^{(1)}\,(0,1) \\
&\quad \oplus M_{15}^{(2)}\,(2,1) \oplus M_6^{(2)}\,(0,2) \oplus M_3^{(2)}\,(1,0) \oplus M_8^{(3)}\,(1,1)\,, \\
M_{10}\,(2,0,0) &= M_6^{(0)}\,(2,0) \oplus M_3^{(1)}\,(1,0) \oplus M_1^{(2)}\,(0,0)\,, \\
M_{20}\,(0,2,0) &= M_6^{(0)}\,(0,2) \oplus M_8^{(1)}\,(1,1) \oplus M_6^{(2)}\,(2,0)\,, \\
M_{60}\,(1,2,0) &= M_{15}^{(0)}\,(1,2) \oplus M_{15}^{(1)}\,(2,1) \oplus M_6^{(1)}\,(0,2) \\
&\quad \oplus M_{10}^{(2)}\,(3,0) \oplus M_8^{(2)}\,(1,1) \oplus M_6^{(3)}\,(2,0)\,, \\
M_{20}\,(3,0,0) &= M_{10}^{(0)}\,(3,0) \oplus M_6^{(1)}\,(2,0) \oplus M_3^{(2)}\,(1,0) \oplus M_1^{(3)}\,(0,0)\,.
\end{aligned}
$$

The decompositions of Figure 5.3.6 appear in Table 5.3.1.

The formula for the **dimension $d(p_1,p_2,p_3)$ of $su(4)$-multiplets** reads (see Section 6.2):

$$
\begin{aligned}
d\,(p_1,p_2p_3) = \tfrac{1}{12}\,&(p_1+1)\,(p_2+1)\,(p_3+1)\,(p_1+p_1+2) \\
&\cdot (p_2+p_3+2)\,(p_1+p_2+p_3+3)\,.
\end{aligned}
\tag{5.14}
$$

The subscript numbers in Figures 5.3.3 up to 5.3.5 are in accordance with d in (5.14). This formula shows that the dimension is independent of an interchange of p_1 and p_3. Because this operation performs only a rotation of the $su(4)$-multiplet (see the remark in connection with Figure 5.3.3), this invariance is expected. On the other hand, interchanges (p_1p_2) or (p_2p_3) influence the dimension d.

The dimension can also be determined by the method of Young tableaux (see at the end of Section 4.6). As an example, we take the multiplet $M_6(0,1,0)$. The diagram consists of a column of two boxes where the numbers 1 up to 4 have to be filled in, following the standard prescriptions. We obtain

$$
\begin{matrix} 1 & 1 & 1 & 2 & 2 & 3 \\ 2 & 3 & 4 & 3 & 4 & 4 \end{matrix}\,. \tag{5.15}
$$

The six pairs stand for dimension 6. In the literature multiplets $M_d\,(p_1,p_2,p_3)$ often are denoted by $[d]$. Because it happens that there are several higher multiplets with the same dimension d, this symbol is ambiguous. We do not use it in this publication.

5.4 The charm C

In Section 4.9 we have seen how the T-, U- and V-axes are replaced by the orthogonal axes T and Y. The representation of $su(4)$-multiplets is three-dimensional, and it should be sufficient to give the distance of the lattice points to the T-Y-plane instead of the W-, X- and Z-values. This "height" is denoted by C, and in particle physics it is named charm. We claim that the coordinate C is the eigenvalue of the operator

$$\boldsymbol{C} = \frac{1}{4}\boldsymbol{1} + \frac{\sqrt{6}}{4}\boldsymbol{\lambda}_{15} = \frac{1}{4}\boldsymbol{1} - \frac{1}{2}\left(\boldsymbol{W}_3 + \boldsymbol{X}_3 + \boldsymbol{Z}_3\right). \tag{5.16}$$

For the last term in (5.16) we have used the definitions (5.5). From Tables 4.2.1 and 5.1.2 we take

$$[\boldsymbol{C}, \boldsymbol{A}] = 0 \quad \text{with} \quad \boldsymbol{A} = \boldsymbol{T}_3,\ \boldsymbol{U}_3,\ \boldsymbol{V}_3,\ \boldsymbol{W}_3,\ \boldsymbol{X}_3 \text{ or } \boldsymbol{Z}_3. \tag{5.17}$$

Therefore, the following eigenvalue relation holds:

$$\boldsymbol{C}\,|T_3U_3V_3W_3X_3Z_3\rangle = C\,|T_3U_3V_3W_3X_3Z_3\rangle\,. \tag{5.18}$$

Analogously, we find

$$[\boldsymbol{C}, \boldsymbol{T}_\pm] = 0 \tag{5.19}$$

and deduce

$$\begin{aligned} \boldsymbol{C}\left(\boldsymbol{T}_\pm\,|T_3U_3V_3W_3X_3Z_3\rangle\right) &= \boldsymbol{T}_\pm\boldsymbol{C}\,|T_3U_3V_3W_3X_3Z_3\rangle \\ &= C\left(\boldsymbol{T}_\pm\,|T_3U_3V_3W_3X_3Z_3\rangle\right). \end{aligned} \tag{5.20}$$

This means, the C-eigenvalue does not change during a step in the T-direction. The same is true for the U- and the V-direction. That is, the C-axis is orthogonal to the T-Y-plane (= T-U-V-plane). By means of Tables 4.2.1 and 5.1.2, we find

$$\begin{aligned} &[\boldsymbol{C},\mathrm{W}_\pm] = \mp\tfrac{1}{2}\cdot 2\,\boldsymbol{W}_\pm \quad \text{and} \\ &\boldsymbol{C}\left(\boldsymbol{W}_\pm\,|T_3U_3V_3W_3X_3Z_3\rangle\right) \\ &\quad = \boldsymbol{W}_\pm\boldsymbol{C}\,|T_3U_3V_3W_3X_3Z_3\rangle \mp \boldsymbol{W}_\pm\,|T_3U_3V_3W_3X_3Z_3\rangle \\ &\quad = (C\mp 1)\left(\boldsymbol{W}_\pm\,|T_3U_3V_3W_3X_3Z_3\rangle\right). \end{aligned} \tag{5.21}$$

We observe that the C-value of a $su(4)$-state decreases by 1 if $\boldsymbol{W}_+$ acts on it. The same is true for $\boldsymbol{X}_+$ and $\boldsymbol{Z}_+$. That is, if we descend from the peak of the multiplet $M_4\,(1,0,0)$ (see Figure 5.3.3) along the edges by a unit step, the height C is also reduced by a unit, which is geometrically impossible. We circumvent the problem by taking the C-values in units of $\sqrt{2/3}$ in analogy to the gauge of the Y-axis. As an example, we specify the state points of the multiplet $M_6\,(0,1,0)$ (see Figure 5.3.3) in T-Y-C-coordinates as follows:

$$\left(\frac{1}{2},-\frac{1}{3},0\right),\left(-\frac{1}{2},-\frac{1}{3},0\right),\left(0,\frac{2}{3},0\right),\left(-\frac{1}{2},\frac{1}{3},1\right),\left(\frac{1}{2},\frac{1}{3},1\right),\left(0,-\frac{2}{3},1\right). \tag{5.22}$$

5.5 Direct products of $su(4)$-multiplets

As in Section 4.13, we do not develop the mathematical formalism of the direct product of multiplets but we accept the graphical method without proof.

In analogy to the procedure for $su(3)$-multiplets, we have to set the second multiplet with its reference point (we choose the centre of the basal plane) in every point of the first multiplet, which we delete afterwards. Clearly this method becomes mostly very confused. Therefore we decompose the $su(4)$-factor-multiplets as in Table 5.3.1 in $su(3)$-multiplets with increasing C-values. The direct product of both sums yields a sum of direct products of $su(3)$-multiplets with equal or different C-values. To every partial direct product we assign the C-value which is the sum of the C-values of both factors. Now we decompose these direct products using Table 4.13.1 or the corresponding technique and attribute the new C-value to the summands. The entire sum of $su(3)$-multiplets again is grouped in $su(4)$ multiplets using Table 5.3.1. This technique is completely equivalent to the original graphical method.

Following this line we handle two examples:

$$\begin{aligned}
&M_4\,(1,0,0) \otimes M_4\,(1,0,0) = \\
&\left[M_3^{(0)}\,(1,0) \oplus M_1^{(1)}\,(0,0)\right] \otimes \left[M_3^{(0)}\,(1,0) \oplus M_1^{(1)}\,(0,0)\right] = \\
&M_3^{(0)}\,(1,0) \otimes M_3^{(0)}\,(1,0) \oplus M_3^{(0)}\,(1,0) \otimes M_1^{(1)}\,(0,0) \oplus \\
&M_1^{(1)}\,(0,0) \otimes M_3^{(0)}\,(1,0) \oplus M_1^{(1)}\,(0,0) \otimes M_1^{(1)}\,(0,0) \triangleq \\
&M_6^{(0)}\,(2,0) \oplus M_3^{(0)}\,(0,1) \oplus \\
&M_3^{(1)}\,(1,0) \oplus M_3^{(1)}\,(1,0) \oplus M_1^{(2)}\,(0,0) = \\
&M_6^{(0)}\,(2,0) \oplus M_3^{(1)}\,(1,0) \oplus M_1^{(2)}\,(0,0) \oplus \\
&M_3^{(0)}\,(0,1) \oplus M_3^{(1)}\,(1,0) = \\
&M_{10}\,(2,0,0) \oplus M_6\,(0,1,0)\,,
\end{aligned} \tag{5.23}$$

$$\begin{aligned}
&M_4\,(1,0,0) \otimes M_4\,(0,0,1) = \\
&\left[M_3^{(0)}\,(1,0) \oplus M_1^{(1)}\,(0,0)\right] \otimes \left[M_1^{(0)}\,(0,0) \oplus M_3^{(1)}\,(0,1)\right] = \\
&M_3^{(0)}\,(1,0) \otimes M_1^{(0)}\,(0,0) \oplus M_3^{(0)}\,(1,0) \otimes M_3^{(1)}\,(0,1) \oplus \\
&M_1^{(1)}\,(0,0) \otimes M_1^{(0)}\,(0,0) \oplus M_1^{(1)}\,(0,0) \otimes M_3^{(1)}\,(0,1) \triangleq \\
&M_3^{(0)}\,(1,0) \oplus M_8^{(1)}\,(1,1) \oplus M_1^{(1)}\,(0,0) \oplus \\
&M_1^{(1)}\,(0,0) \oplus M_3^{(2)}\,(0,1) = \\
&M_{15}\,(1,0,1) \oplus M_1^{(1)}\,(0,0,0)\,.
\end{aligned} \tag{5.24}$$

in the same way one derives

$$\begin{aligned}
&M_4\,(1,0,0) \otimes M_4\,(1,0,0) \otimes M_4\,(1,0,0) \triangleq \\
&M_{20}\,(3,0,0) \oplus 2 \cdot M_{20}\,(1,1,0) \oplus M_4\,(0,0,1)\,.
\end{aligned} \tag{5.25}$$

In Section 4.14 the decomposition of direct products of multiplets is treated by Young diagrams. In Figure 5.5.1 we apply it to the case (5.24):

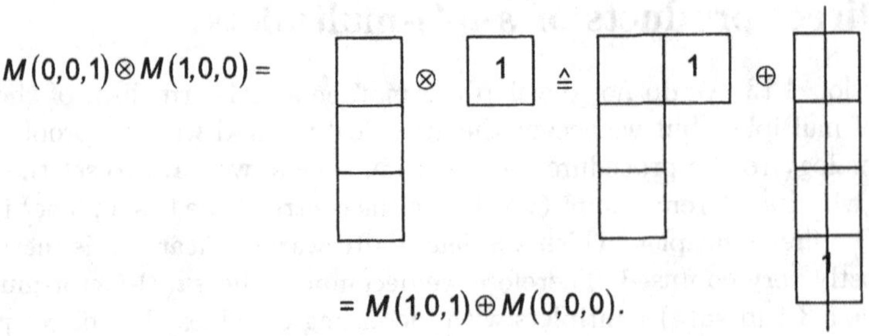

Figure 5.5.1. Direct product $M\,(0,0,1) \otimes M\,(1,0,0)$.

In Table 5.5.1 further examples are given.

Table 5.5.1. Decompositions of direct products of small $su(4)$-multiplets.

$$
\begin{aligned}
&M_4\,(1,0,0) \otimes M_4\,(1,0,0) \triangleq M_{10}\,(2,0,0) \oplus M_6\,(0,1,0)\,,\\
&M_4\,(1,0,0) \otimes M_4\,(0,0,1) \triangleq M_{15}\,(1,0,1) \oplus M_1\,(0,0,0)\,,\\
&M_6\,(0,1,0) \otimes M_4\,(1,0,0) \triangleq M_{20}\,(1,1,0) \oplus M_4\,(0,0,1)\,,\\
&M_{10}\,(2,0,0) \otimes M_4\,(1,0,0) \triangleq M_{20}\,(3,0,0) \oplus M_{20}\,(1,1,0)\,,\\
&M_{10}\,(0,0,2) \otimes M_4\,(1,0,0) \triangleq M_{36}\,(1,0,2) \oplus M_4\,(0,0,1)\,,\\
&M_{15}\,(1,0,1) \otimes M_4\,(1,0,0) \triangleq M_{36}\,(2,0,1) \oplus M_{20}\,(0,1,1) \oplus M_4\,(1,0,0)\,,\\
&M_{20}\,(3,0,0) \otimes M_4\,(1,0,0) \triangleq M_{35}\,(4,0,0) \oplus M_{45}\,(2,1,0)\,,\\
&M_{20}\,(0,0,3) \otimes M_4\,(1,0,0) \triangleq M_{70}\,(1,0,3) \oplus M_{10}\,(0,0,2)\,,\\
&M_{20}\,(1,1,0) \otimes M_4\,(1,0,0) \triangleq M_{45}\,(2,1,0) \oplus M_{20}\,(0,2,0) \oplus M_{15}\,(1,0,1)\,,\\
&M_{20}\,(0,1,1) \otimes M_4\,(1,0,0) \triangleq M_{64}\,(1,1,1) \oplus M_{10}\,(0,0,2) \oplus M_6\,(0,1,0)\,,\\
&M_{20}\,(0,2,0) \otimes M_4\,(1,0,0) \triangleq M_{60}\,(1,2,0) \oplus M_{20}\,(0,1,1)\,,\\
&M_{20}\,(1,1,0) \otimes M_{15}\,(1,0,1) \triangleq M_{140}\,(2,1,1) \oplus M_{20}\,(3,0,0) \oplus M_{60}\,(0,2,1)\\
&\qquad\qquad \oplus 2 \cdot M_{20}\,(1,1,0) \oplus M_{36}\,(1,2,0) \oplus M_4\,(0,0,1)\,.
\end{aligned}
$$

5.6 The Cartan–Weyl basis of $su(4)$

In this section further structural properties of the $su(4)$-algebra are put together. Some relations are given without proof, but a lot of checks using the commutator tables are made. Most of the rules discussed hold also for other $su(N)$-algebras or even more generally for semisimple Lie algebras.

We start with the **weight operators** $\boldsymbol{T}_3$, $\boldsymbol{Y}$ (see Section 4.9) and a variant on $\boldsymbol{C}$ (Section 5.4), which we denote by $\boldsymbol{H}_1$, $\boldsymbol{H}_2$ and $\boldsymbol{H}_3$:

$$\begin{aligned} \frac{1}{2}\boldsymbol{\lambda}_3 &= \boldsymbol{T}_3 = \boldsymbol{H}_1, \\ \frac{1}{\sqrt{3}}\boldsymbol{\lambda}_8 &= \boldsymbol{Y} = \frac{2}{3}(\boldsymbol{U}_3 + \boldsymbol{V}_3) = \boldsymbol{H}_2, \\ \frac{\sqrt{6}}{2}\boldsymbol{\lambda}_{15} &= \boldsymbol{W}_3 + \boldsymbol{X}_3 + \boldsymbol{Z}_3 = \boldsymbol{H}_3. \end{aligned} \tag{5.26}$$

With the help of (4.13) and (5.6) one finds

$$\begin{aligned} \boldsymbol{T}_3 &= \boldsymbol{H}_1, \\ 2\boldsymbol{U}_3 &= -\boldsymbol{H}_1 + \frac{3}{2}\boldsymbol{H}_2, \\ 2\boldsymbol{V}_3 &= \boldsymbol{H}_1 + \frac{3}{2}\boldsymbol{H}_2, \\ 2\boldsymbol{W}_3 &= \boldsymbol{H}_1 + \frac{1}{2}\boldsymbol{H}_2 + \frac{2}{3}\boldsymbol{H}_3, \\ 2\boldsymbol{X}_3 &= -\boldsymbol{H}_1 + \frac{1}{2}\boldsymbol{H}_2 + \frac{2}{3}\boldsymbol{H}_3, \\ 2\boldsymbol{Z}_3 &= \qquad -\boldsymbol{H}_2 + \frac{2}{3}\boldsymbol{H}_3. \end{aligned} \tag{5.27}$$

From the commutator Tables 4.2.1 and 5.1.2 we take

$$[\boldsymbol{H}_i, \boldsymbol{H}_k] = 0, \quad i,k = 1,2,3. \tag{5.28}$$

We rename the step operators like this:

$$\begin{aligned} \boldsymbol{T}_\pm &= \tfrac{1}{2}(\boldsymbol{\lambda}_1 \pm \mathrm{i}\boldsymbol{\lambda}_2) = \boldsymbol{E}_{\pm 1}, \\ \boldsymbol{V}_\pm &= \tfrac{1}{2}(\boldsymbol{\lambda}_4 \pm \mathrm{i}\boldsymbol{\lambda}_5) = \boldsymbol{E}_{\pm 2}, \\ \boldsymbol{U}_\pm &= \tfrac{1}{2}(\boldsymbol{\lambda}_6 \pm \mathrm{i}\boldsymbol{\lambda}_7) = \boldsymbol{E}_{\pm 3}, \\ \boldsymbol{W}_\pm &= \tfrac{1}{2}(\boldsymbol{\lambda}_9 \pm \mathrm{i}\boldsymbol{\lambda}_{10}) = \boldsymbol{E}_{\pm 4}, \\ \boldsymbol{X}_\pm &= \tfrac{1}{2}(\boldsymbol{\lambda}_{11} \pm \mathrm{i}\boldsymbol{\lambda}_{12}) = \boldsymbol{E}_{\pm 5}, \\ \boldsymbol{Z}_\pm &= \tfrac{1}{2}(\boldsymbol{\lambda}_{13} \pm \mathrm{i}\boldsymbol{\lambda}_{14}) = \boldsymbol{E}_{\pm 6}. \end{aligned} \tag{5.29}$$

The operators $\boldsymbol{H}_1$ up to $\boldsymbol{H}_3$ and $\boldsymbol{E}_{\pm 1}$ up to $\boldsymbol{E}_{\pm 6}$ are named *Cartan–Weyl basis* of $su(4)$.

Now we look on **commutators of $\boldsymbol{H}$-operators with $\boldsymbol{E}$-operators** and take the example

$$\begin{aligned} &[\boldsymbol{H}_3, \boldsymbol{E}_{+5}] = [\boldsymbol{W}_3, \boldsymbol{X}_+] + [\boldsymbol{X}_3, \boldsymbol{X}_+] + [\boldsymbol{Z}_3, \boldsymbol{X}_+] = \left(\tfrac{1}{2} + 1 + \tfrac{1}{2}\right)\boldsymbol{X}_+ \\ &= 2\boldsymbol{E}_{+5} \equiv \alpha_{3;5}\boldsymbol{E}_{+5} \qquad \text{with } \alpha_{3;5} = 2. \end{aligned} \tag{5.30}$$

All commutators of this type yield similar results as follows:

$$[\boldsymbol{H}_i, \boldsymbol{E}_k] = \alpha_{ik}\boldsymbol{E}_k, \qquad i = 1,2,3, \quad k = \pm 1, \pm 2, \ldots, \pm 6. \tag{5.31}$$

Table 5.6.1 contains the so-called root components α_{ik} for positive k.

Table 5.6.1. Coefficients α_{ik} for positive k.

k	α_{1k}	α_{2k}	α_{3k}
1	1	0	0
2	1/2	1	0
3	−1/2	1	0
4	1/2	1/3	2
5	−1/2	1/3	2
6	0	−2/3	2

For negative indices k one obtains

$$\alpha_{i,-k} = -\alpha_{ik}. \tag{5.32}$$

The vector $\underline{\alpha}_k = (\alpha_{1k}, \alpha_{2k}, \alpha_{3k})$, $k = \pm 1, \pm 2, \ldots, \pm 6$, is named *root vector.*

Now we investigate **commutators of $\boldsymbol{E}_k$ with $\boldsymbol{E}_{-k}$** and choose

$$[\boldsymbol{E}_5, \boldsymbol{E}_{-5}] = [\boldsymbol{X}_+, \boldsymbol{X}_-] = 2\boldsymbol{X}_3 = -1 \cdot \boldsymbol{H}_1 + \frac{1}{2}H_2 + \frac{2}{3}\boldsymbol{H}_3 \tag{5.33}$$

where (5.27) has been used. We sum up:

$$[\boldsymbol{E}_5, \boldsymbol{E}_{-5}] = \sum_{i=1}^{3} \gamma_{5;i}\, \boldsymbol{H}_i \quad \text{with} \quad \gamma_{5;1} = -1,\ \gamma_{5;2} = \frac{1}{2},\ \gamma_{5;3} = \frac{2}{3}. \tag{5.34}$$

Generally, one obtains for such commutators

$$[\boldsymbol{E}_k, \boldsymbol{E}_{-k}] = \sum_{i=1}^{3} \gamma_{ki}\, \boldsymbol{H}_i\,. \tag{5.35}$$

Table 5.6.2 presents the values of the coefficients γ_{ki} for positive k.

Table 5.6.2. Coefficients γ_{ki}

k	γ_{k1}	γ_{k2}	γ_{k3}
1	2	0	0
2	1	3/2	0
3	−1	3/2	0
4	1	1/2	2/3
5	−1	1/2	2/3
6	0	−1	2/3

Replacing k by $-k$ interchanges the operators in the commutator (5.35), which yields

$$\gamma_{-k,i} = -\gamma_{ki}. \tag{5.36}$$

We define the vector $\underline{\gamma}_k = (\gamma_{k1}, \gamma_{k2}, \gamma_{k3})$.

Now we investigate the **commutator** $[\boldsymbol{E}_k, \boldsymbol{E}_l]$ with $k + l \neq 0$. We claim

$$[\boldsymbol{E}_k, \boldsymbol{E}_l] = N_{kl} \boldsymbol{E}_m \quad \text{with } k + l \neq 0, \tag{5.37}$$

where the indices fulfil the vector condition

$$\underline{\alpha}_k + \underline{\alpha}_l = \underline{\alpha}_m \,. \tag{5.38}$$

Of course, the relation

$$N_{lk} = -N_{kl} \tag{5.39}$$

holds. Due to (5.32), we have

$$\underline{\alpha}_{-i} = -\underline{\alpha}_i, \tag{5.40}$$

and with sign reversed indices k, l, m the equation (5.38) is satisfied as well. We investigate the following examples numerically using Table 5.1.2

$$\begin{aligned} [\boldsymbol{E}_5, \boldsymbol{E}_2] &= [\boldsymbol{X}_+ \boldsymbol{V}_+] = 0, \\ [\boldsymbol{E}_5, \boldsymbol{E}_{-3}] &= [\boldsymbol{X}_+, \boldsymbol{U}_-] = -\boldsymbol{Z}_+ = -\boldsymbol{E}_6 \,, \quad \text{i.e. } N_{5,-3} = -1, \ m = 6, \end{aligned} \tag{5.41}$$

and with the help of Table 5.6.1 we find $\underline{\alpha}_5 + \underline{\alpha}_{-3} = \left(0, -\frac{2}{3}, 2\right) = \underline{\alpha}_6$, which satisfies (5.38). Table 5.6.3 puts together non-vanishing values N_{kl} and the affiliated indices m.

Table 5.6.3. Coefficients N_{kl} for positive k.

k	l	m	N_{kl}	k	l	m	N_{kl}
6	3	5	−1	4	−1	5	−1
6	2	4	−1	4	−2	6	−1
6	−4	−2	1	3	1	2	−1
6	−5	−3	1	3	−2	−1	1
5	1	4	−1	2	−1	3	−1
5	−3	6	−1				
5	−4	−1	1				

Numerical inspection shows

$$N_{-k,-l} = -N_{kl}. \tag{5.42}$$

Making use of (5.39), (5.42) and Table 5.6.3, the remaining non-vanishing coefficients N_{kl} can be found.

The inner product of the root vectors $\underline{\gamma}_k$ and $\underline{\alpha}_l$,

$$\left(\underline{\gamma}_k, \underline{\alpha}_l\right) = \sum_{m=1}^{3} \gamma_{km}\alpha_{ml}, \tag{5.43}$$

contains further laws. Table 5.6.4 presents such products.

Table 5.6.4. Inner products of $\underline{\gamma}_k$ and $\underline{\alpha}_l$, values for $\left(\underline{\gamma}_k, \underline{\alpha}_l\right)$.

$k\backslash\ l=$	1	2	3	4	5	6
1	2	1	−1	1	−1	0
2	1	2	1	1	0	−1
3	−1	1	2	0	1	−1
4	1	1	0	2	1	1
5	−1	0	1	1	2	1
6	0	−1	−1	1	1	2

We make a spot check and calculate $\left(\underline{\gamma}_4, \underline{\alpha}_5\right) = -\frac{1}{2} + \frac{1}{2}\cdot\frac{1}{3} + \frac{2}{3}\cdot 2 = 1$ as given in Table 5.6.4. The table is symmetric with regard to the diagonal. For negative indices the relations

$$\left(\underline{\gamma}_k, \underline{\alpha}_l\right) = -\left(\underline{\gamma}_{-k}, \underline{\alpha}_l\right) = -\left(\underline{\gamma}_k, \underline{\alpha}_{-l}\right) = \left(\underline{\gamma}_{-k}, \underline{\alpha}_{-l}\right) \text{ hold.} \tag{5.44}$$

The expressions

$$2\frac{\left(\underline{\gamma}_k, \underline{\alpha}_l\right)}{\left(\underline{\gamma}_k, \underline{\alpha}_k\right)} \tag{5.45}$$

are particularly important. The values of this expression are identical with the values for $\left(\underline{\gamma}_k, \underline{\alpha}_l\right)$ in Table 5.6.4, because the diagonal contains only 2's. We can state that the expression (5.45) takes only the values 0, ±1, ±2:

$$2\frac{\left(\underline{\gamma}_k, \underline{\alpha}_l\right)}{\left(\underline{\gamma}_k, \underline{\alpha}_k\right)} = 0,\ \pm 1,\ \pm 2\ . \tag{5.46}$$

Taking into account negative indices on the left-hand side does not change the statement. The theory of classical Lie algebras teaches that the expression (5.45) equals one of the values 0, ±1, ±2, ±3:

$$2\frac{\left(\underline{\gamma}_k, \underline{\alpha}_l\right)}{\left(\underline{\gamma}_k, \underline{\alpha}_k\right)} = 0,\ \pm 1,\ \pm 2,\ \pm 3\ . \tag{5.47}$$

Our result (5.46) is in accordance with this rule.

Another **theorem** of the group theory states that, if $\underline{\alpha}_l$ and $\underline{\alpha}_k$ are root vectors of a so-called semisimple Lie algebra, the expression

$$\underline{\alpha}_l - \underline{\alpha}_k \cdot 2\frac{\left(\underline{\gamma}_k, \underline{\alpha}_l\right)}{\left(\underline{\gamma}_k, \underline{\alpha}_k\right)} = \underline{\alpha}_n \tag{5.48}$$

is also a root vector $\underline{\alpha}_n$. We calculate an example with $l = 5$, $k = 3$ making use of Tables 5.6.1 and 5.6.4:

$$\left(-\frac{1}{2}, \frac{1}{3}, 2\right) - \left(-\frac{1}{2}, 1, 0\right) \cdot 1 = \left(0, -\frac{3}{2}, 2\right) = \underline{\alpha}_6, \quad \text{i.e. } n = 6. \tag{5.49}$$

For $l = k$ we obtain $\underline{\alpha}_l - \underline{\alpha}_k \cdot 2 = -\underline{\alpha}_l = \underline{\alpha}_{-l}$, i.e. $n = -l$.

Table 5.6.5 presents the indices $n(l, k)$ defined in (5.48) for positive l and k.

Table 5.6.5. Values $n(l, k)$.

$l\backslash\ k=$	6	5	4	3	2	1
6	−6	−3	−2	5	4	6
5	3	−5	−1	6	5	4
4	2	1	−4	4	6	5
3	5	−6	3	−3	−1	2
2	4	2	−6	1	−2	3
1	1	4	−5	2	−3	−1

Due to (5.40) and (5.36) we have

$$2\frac{\left(\underline{\gamma}_{-k}, \underline{\alpha}_l\right)}{\left(\underline{\gamma}_{-k}, \underline{\alpha}_{-k}\right)} = -2\frac{\left(\underline{\gamma}_k, \underline{\alpha}_l\right)}{\left(\underline{\gamma}_k, \underline{\alpha}_k\right)} \quad \text{and } n\left(l, -k\right) = n\left(l, k\right). \tag{5.50}$$

In addition we obtain

$$\underline{\alpha}_{-l} - \underline{\alpha}_k \cdot 2\frac{\left(\underline{\gamma}_k, \underline{\alpha}_{-l}\right)}{\left(\underline{\gamma}_k, \underline{\alpha}_k\right)} = -\left[\underline{\alpha}_l - \underline{\alpha}_k \cdot 2\frac{\left(\underline{\gamma}_k, \underline{\alpha}_l\right)}{\left(\underline{\gamma}_k, \underline{\alpha}_k\right)}\right], \quad \text{i.e., } n\left(-l, k\right) = -n\left(l, k\right). \tag{5.51}$$

By means of (5.50) and (5.51), we can extend Table 5.6.5 to negative values for l and k.

In the next step, we investigate the **normalisation of the coefficients** N_{kl} (see (5.39)). In (5.37) instead of the indices k, l and m the index notation $\underline{\alpha}_k$, $\underline{\alpha}_l$ and $\underline{\alpha}_k + \underline{\alpha}_l$ often is used and the relations (5.37) and (5.38) read

$$\left[\boldsymbol{E}_{\underline{\alpha}_k}, \boldsymbol{E}_{\underline{\alpha}_l}\right] = N_{\underline{\alpha}_k\underline{\alpha}_l} \boldsymbol{E}_{\underline{\alpha}_k + \underline{\alpha}_l}. \tag{5.52}$$

If $\underline{\alpha}_k + \underline{\alpha}_l$ is not a root, the right hand side of (5.52) vanishes.

We now assume that $N_{\underline{\alpha}_k\underline{\alpha}_l}$ does not vanish and let follow a similar step $\left[\boldsymbol{E}_{\underline{\alpha}_k}, \boldsymbol{E}_{\underline{\alpha}_k+\underline{\alpha}_l}\right] = N_{\underline{\alpha}_k,\underline{\alpha}_k+\underline{\alpha}_l}\boldsymbol{E}_{2\underline{\alpha}_k+\underline{\alpha}_l}$, which contains another root, $2\underline{\alpha}_k + \underline{\alpha}_l$, supposed it does not vanish. Because the number of roots is limited, this procedure, which generates a root series based on $\underline{\alpha}_k$ and containing $\underline{\alpha}_l$, must stop once. Let $n^+_{kl}\underline{\alpha}_k + \underline{\alpha}_l$, n^+_{kl} integer, be the last non-vanishing root. In the same way the root series

$$\underline{\alpha}_l\,, \quad \underline{\alpha}_l - \underline{\alpha}_k, \quad \underline{\alpha}_l - 2\underline{\alpha}_k, \quad \ldots, \quad \underline{\alpha}_l - n^-_{kl}\underline{\alpha}_k \tag{5.53}$$

is generated. Group theory states the following relation:

$$N_{\underline{\alpha}_k\underline{\alpha}_l} \cdot N_{-\underline{\alpha}_k,\underline{\alpha}_k+\underline{\alpha}_l} = \frac{n^+_{kl}}{2}\left(n^-_{kl}+1\right)\left(\gamma_k, \alpha_k\right). \tag{5.54}$$

In order to check it with our $su(4)$-data, we look for n^+_{kl} and n^-_{kl} for the pair (k,l) referred to in Table 5.6.3. For example, we take $\underline{\alpha}_5 + \underline{\alpha}_1 = \underline{\alpha}_4$ and $2\underline{\alpha}_5 + \underline{\alpha}_1 = \underline{\alpha}_5 + \underline{\alpha}_4 = \left(0, \frac{2}{3}, 4\right)$, which is not a root (see Table 5.6.1). Therefore, we have $n^+_{5;1} = 1$. Analogously, we obtain $-\underline{\alpha}_5 + \underline{\alpha}_1 = \left(\frac{3}{2}, -\frac{1}{3}, -2\right)$, which is not a root as well and we get $n^-_{5;1} = 0$. The same n-values appear for all pairs (k,l):

$$n^+_{kl} = 1, \quad n^-_{kl} = 0, \tag{5.55}$$

which we insert in (5.54) using the diagonal values in Table 5.6.4:

$$N_{kl} \cdot N_{-k,m(k,l)} = \frac{1}{2} \cdot 1 \cdot 2 = 1. \tag{5.56}$$

The value $m(k,l)$ is associated to k and l according to Table 5.6.3.

We inspect our N-values in Table 5.6.3. For instance we take $k = 4$ and $l = -2$: $N_{4,-2} \cdot N_{-4,m(4,-2)=6} = N_{4,-2} \cdot (-N_{6,-4}) = (-1)\,(-1) = 1$ in accordance with (5.56). For all pairs (k,l) the relation (5.56) is satisfied.

The equations (5.28), (5.31), (5.35) and (5.37), (5.38) are named *canonical relations.*

The results apply also to $su(3)$ if we drop $\boldsymbol{H}_3$ and $\boldsymbol{E}_{\pm 4}$ up to $\boldsymbol{E}_{\pm 6}$. If we leave out $\boldsymbol{H}_2$, too, and also $\boldsymbol{E}_{\pm 2}$ and $\boldsymbol{E}_{\pm 3}$, this theory is applied to $su(2)$. The laws of this section hold not only for $su(N)$-algebras but also correspondignly for the so-called semisimple Lie algebras and result in a systematic order.

Chapter 6

General properties of the $su(N)$-algebras

In this chapter above all, properties of the $su(N)$-algebras (with arbitrary N), which have been mentioned in the previous chapters, are put together and completed.

The generators are formulated generally. The recursion formula for the dimensions of $su(N)$-multiplets is applied.

For fundamental multiplets the eigenvalues of the quadratic Casimir operator are developed and several direct products of multiplets are dealt with.

6.1 Elements of the $su(N)$-algebra

The basis elements of the real $su(N)$-matrix-algebra with vanishing traces are the anti-Hermitian matrices $\underline{\underline{e}}_i = -\frac{1}{2}\mathrm{i}\underline{\underline{\lambda}}_i$, (2.15). The Hermitian $N \times N$-matrices $\underline{\underline{\lambda}}_i$ (generators) are given partly in (2.5). For the remaining diagonal matrices usually one does not choose the obvious form (2.6) but one takes the set

$$\begin{pmatrix} 1 & & & & & \\ & -1 & & & 0 & \\ & & 0 & & & \\ & & & 0 & & \\ & 0 & & & \ddots & \\ & & & & & 0 \end{pmatrix}, \quad \frac{1}{\sqrt{3}}\begin{pmatrix} 1 & & & & & \\ & 1 & & & 0 & \\ & & -2 & & & \\ & & & 0 & & \\ & 0 & & & \ddots & \\ & & & & & 0 \end{pmatrix},$$

$$\frac{1}{\sqrt{6}}\begin{pmatrix} 1 & & & & & & \\ & 1 & & & 0 & & \\ & & 1 & & & & \\ & & & -3 & & & \\ & & & & 0 & & \\ & 0 & & & & \ddots & \\ & & & & & & 0 \end{pmatrix}, \tag{6.1}$$

$$\ldots, \sqrt{\frac{2}{N(N-1)}}\begin{pmatrix} 1 & & & & & \\ & 1 & & & 0 & \\ & & 1 & & & \\ & & & \ddots & & \\ & & & & \ddots & \\ & 0 & & & 1 & \\ & & & & & -N+1 \end{pmatrix}$$

(see (4.2) and (5.3)). This set of diagonal matrices causes the trace condition (2.27) to be satisfied for all matrices $\underset{=}{\lambda}_i$ and $\underset{=}{\lambda}_l$. This property produces the total antisymmetry of the structure constants C_{ikl} (see (2.30)). The set (6.1) contains $N-1$ generators which mutually commute, that is, the Lie algebra $su(N)$ has the rank $N-1$ (see 5.4) . Therefore, there are $N-1$ Casimir operators (Section 4.11). The generators (6.1) turn into the weight operators $\boldsymbol{H}_1$, $\boldsymbol{H}_2, \ldots, \boldsymbol{H}_{N-1}$ (see (5.26)). In order to determine the structure constants of $su(N)$, no closed formulas are known, they have to be calculated by means of (2.29) performing matrix multiplications.

The Casimir operators of $su(N)$ were discussed in Section 4.11, and an expression for the quadratic Casimir operator $\boldsymbol{C}_1$ was given in (4.77).

6.2 Multiplets of $su(N)$

Acting on functions of a multiplet, the $N-1$ Casimir operators generate $N-1$ eigenvalues and finally $N-1$ parameters p_1, $p_2, \ldots, p_{N-1}$, which characterise the multiplet. They define the structure of the multiplet and lay down its dimension $d\,(p_1, p_2, \ldots, p_{N-1})$. It can be shown that the general formula for d is insensitive to an interchange of p_1 and p_{N-1}. We have discussed this property for $su(4)$ in connection with (5.14).

There is a **recursion formula** which relates the dimension $d_{N-1}\,(p_1, p_2, \ldots, p_{N-2})$ of a $su(N-1)$-multiplet to the dimension $d_N\,(p_1, p_2, \ldots, p_{N-1})$ of a $su(N)$-multiplet:

$$\begin{array}{l} d_N\,(p_1, p_2, \ldots, p_{N-1}) = \\ \frac{1}{(N-1)!}\,(p_{N-1}+1)\,(p_{N-1}+p_{N-2}+2)\,(p_{N-1}+p_{N-2}+p_{N-3}+3)\cdot \\ \ldots\cdot(p_{N-1}+\ldots+p_1+N-1)\cdot d_{N-1}\,(p_1, p_2, \ldots, p_{N-2})\,. \end{array} \tag{6.2}$$

For $N=2$ we use $d_1 = 1$ and obtain

$$d_2\,(p_1) = p_1 + 1, \tag{6.3}$$

which coincides with $2j+1$ (see (4.27)). For $N = 3$ we get

$$d_3\,(p_1, p_2) = \frac{1}{2}\,(p_2+1)\,(p_2+p_1+2)\,(p_1+1) \tag{6.4}$$

in accordance with (4.56). The next analogue step yields

$$\begin{array}{rl} d_4\,(p_1, p_2, p_3) = & \frac{1}{6}\,(p_3+1)\,(p_3+p_2+2)\,(p_3+p_2+p_1+3) \\ & \cdot\frac{1}{2}\,(p_2+1)\,(p_2+p_1+2)\,(p_1+1) \end{array} \tag{6.5}$$

in accordance with (5.14).

The recursion formalism (6.2) is very suitable for computer programs and the existing more compact method (Coleman, 1968) nowadays seems not to be necessary anymore.

Up to the end of this section we will deal with **fundamental multiplets**. The multiplet $M(1,0,\ \ldots\ ,0)$ (with $N-2$ zeros) is named fundamental multiplet for $su(N)$ (see Section 4.8 and (2.19)). The expressions (6.3) up to (6.5) yield $d_N = N$ for these multiplets, which is true for all N, because, if we insert $d_{N-1}\,(1, 0, \ldots, 0) = N-1$ in (6.2) we obtain

$$d_N\,(1, 0, \ldots, 0) = \frac{1}{(N-1)!}1\cdot 2\cdot\ldots\cdot(N-2)\,N\,(N-1) = N. \tag{6.6}$$

Due to (2.19), the fundamental multiplet $\psi_1,\ \psi_2, \ldots, \psi_N$ meets

$$\boldsymbol{e}_j\psi_k = \sum_{l-1}^{N} (e_j)_{lk}\psi_l \tag{6.7}$$

and with $\boldsymbol{e}_j = -2\mathrm{i}\boldsymbol{\lambda}_j$ and $\underline{\underline{e}}_j = -2\mathrm{i}\underline{\underline{\lambda}}_j$ we obtain

$$\boldsymbol{\lambda}_i\psi_k = \sum_{l=1}^{N} (\lambda_j)_{lk}\psi_l \quad \text{(fundamental multiplet)}\,. \tag{6.8}$$

Now, we apply the **quadratic Casimir operator** $\boldsymbol{C}_1$ on ψ_k of the fundamental multiplet. Due to (4.77) and (4.82) we write $\boldsymbol{C}_1 = \frac{1}{4}\sum_{i=1}^{n} \boldsymbol{\lambda}_i^2$, where $n = N^2 - 1$ is the dimension of the algebra $su(N)$.

$$
\begin{aligned}
\boldsymbol{C}_1\psi_k &= \frac{1}{4}\sum_{i=1}^{n} \boldsymbol{\lambda}_i(\boldsymbol{\lambda}_i\psi_k) = \frac{1}{4}\sum_{i=1}^{n} \boldsymbol{\lambda}_i \sum_{l=1}^{N} (\lambda_i)_{lk}\psi_l \qquad (6.9)\\
&= \frac{1}{4}\sum_{i=1}^{n}\sum_{l=1}^{N} (\lambda_i)_{lk} \sum_{m=1}^{N} (\lambda_i)_{ml}\psi_m = \sum_{m=1}^{N} \frac{1}{4}\left(\sum_{i=1}^{n} \underline{\underline{\lambda}}_i^2\right)_{mk} \psi_m .
\end{aligned}
$$

Next we assert that $\sum_{i=1}^{n} \underline{\underline{\lambda}}_i^2$ is proportional to the unit matrix $\underline{\underline{1}}_N$. For example, in $su(3)$ we have

$$
\begin{aligned}
\underline{\underline{\lambda}}_1^2 = \underline{\underline{\lambda}}_2^2 = \underline{\underline{\lambda}}_3^2 &= \begin{pmatrix} 1 & 0 & 0 \\ 0 & 1 & 0 \\ 0 & 0 & 0 \end{pmatrix}, \underline{\underline{\lambda}}_4^2 = \underline{\underline{\lambda}}_5^2 = \begin{pmatrix} 1 & 0 & 0 \\ 0 & 0 & 0 \\ 0 & 0 & 1 \end{pmatrix},\\
\underline{\underline{\lambda}}_6^2 = \underline{\underline{\lambda}}_7^2 &= \begin{pmatrix} 0 & 0 & 0 \\ 0 & 1 & 0 \\ 0 & 0 & 1 \end{pmatrix}, \ \underline{\underline{\lambda}}_8^2 = \frac{1}{3}\begin{pmatrix} 1 & 0 & 0 \\ 0 & 1 & 0 \\ 0 & 0 & 4 \end{pmatrix} \qquad (6.10)
\end{aligned}
$$

and we obtain

$$
\frac{1}{4}\sum_{i=1}^{n} \underline{\underline{\lambda}}_i^2 = \frac{4}{3}\underline{\underline{1}}_3 \quad (su\,(3))\,. \qquad (6.11)
$$

Generally, this expression reads

$$
\frac{1}{4}\sum_{i=1}^{n=N^2-1} \underline{\underline{\lambda}}_i^2 = \frac{N^2-1}{2N}\,\underline{\underline{1}}_N \qquad (6.12)
$$

(see Greiner, 1994, p. 376). We insert it in (6.9) as follows:

$$
\boldsymbol{C}_1\psi_k = \frac{N^2-1}{2N}\sum_{m=1}^{N} \delta_{mk}\psi_m = \frac{N^2-1}{2N}\psi_k \quad \text{(fundamental multiplet)}\,. \qquad (6.13)
$$

We observe that the eigenvalue of the quadratic Casimir operator for a fundamental multiplet is $\frac{N^2-1}{2N}$ — of course independent of the properties of the individual functions.

Direct products of multiplets with $N \geq 4$ are intricate. For the decomposition in a Clebsch–Gordan series the Young diagram technique is recommended (Sections 4.14, 5.5). By means of Figure 6.2.1, we investigate the direct product of two identical fundamental multiplets:

$$M(1,0,\ldots,0) \otimes M(1,0,\ldots,0) = \quad \boxed{} \otimes \boxed{1} \triangleq \boxed{\;\;1} \oplus \boxed{\begin{array}{c}\\\hline 1\end{array}}$$

$$\underset{N-2}{\longleftrightarrow} \qquad \underset{N-2}{\longleftrightarrow}$$

$$= M(2,0,\ldots,0) \oplus M(0,1,0,\ldots,0).$$

$$\underset{N-2}{\longleftrightarrow} \qquad \underset{N-3}{\longleftrightarrow}$$

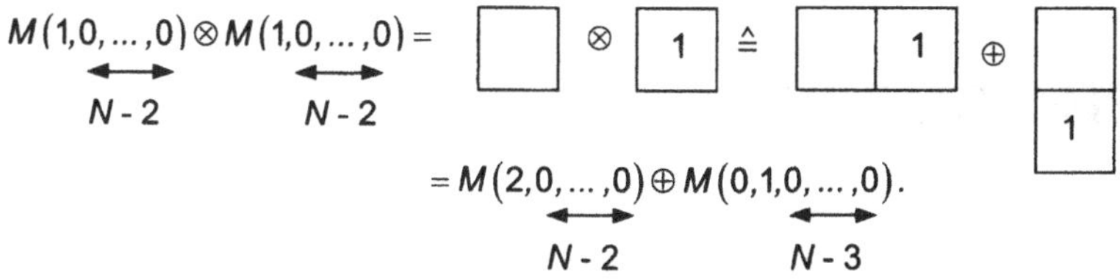

Figure 6.2.1. Direct product $M(1,0,\ldots,0) \otimes M(1,0,\ldots,0)$.

We determine the dimension of the multiplet $M(2,0,\ldots,0)$ and claim $d_N(2,0,\ldots,0) = N(N+1)/2$, which is true particularly for $su(3)$ (use (4.56)). For the general case we insert $d_{N-1}(2,0,\ldots,0) = N(N-1)/2$ in (6.2) and obtain

$$
\begin{aligned}
d_N(2,0,\ldots,0) &= \frac{1}{(N-1)!} 1 \cdot 2 \cdot \ldots \cdot (N-2)(N+1) \cdot N(N-1)/2 \\
&= N(N+1)/2, \qquad \text{QED.} \qquad\qquad (6.14)
\end{aligned}
$$

Analogously, one shows $d_N(0,1,0,\ldots,0) = N(N-1)/2$. In the relation of Figure 6.2.1 we put in the dimensions as subscripts:

$$M_N(1,0,\ldots,0) \otimes M_N(1,0,\ldots,0)$$
$$\triangleq M_{N(N+1)/2}(2,0,\ldots,0) \oplus M_{N(N-1)/2}(0,1,0,\ldots,0). \qquad (6.15)$$

The dimension check $N^2 = N(N+1)/2 + N(N-1)/2$ is satisfied. Analogously, for a fundamental multiplet and an anti-multiplet the relation

$$M_N(1,0,\ldots,0) \otimes M_N(0,\ldots0,1) \triangleq M_{N^2-1}(1,0,\ldots,0,1) \oplus M_1(0,0,\ldots,0) \quad (6.16)$$

holds.

References

Biedenharn, L. C., Louck, J. D., Carruthers, P. A. (1981), *Angular Momentum in Quantum Physics*, Addison-Wesley Publishing Company, Reading.

Brussaard. P. J. and Glaudemans, P. W. M. (1977), *Shell Model Applications in Nuclear Spectroscopy*, North Holland, Amsterdam.

Carter, R., Segal, G., Macdonald, I. (1995), *Lectures on Lie Groups and Lie Algebras*, Cambridge University Press.

Coleman, H. J. (1968), *Symmetry Groups made easy*, Adv. Quantum Chemistry 4, 83.

Cornwell, J. F. (1990), *Group Theory in Physics*, Academic Press, London.

Edmonds, A. (1996), *Angular Momentum in Quantum Mechanics*, Princeton Landmarks in Physics, Princeton.

Fuchs, J., Schweigert, Ch. (1997), *Symmetries, Lie Algebras and Representations*, Cambridge Universitiy Press.

Fulton, W., Harris J. (1991), *Representation Theory, A First Course*, Springer, New York.

Gell-Mann, M., Ne'eman, Y. (1964), *The Eight-fold Way*, Benjamin, New York.

Greiner, W., Müller, B. (1994), *Quantum Mechanics, Symmetries*, Springer, Berlin.

Lucha, W., Schöberl, F. F. (1993), *Gruppentheorie, Eine elementare Einführung für Physiker*, BI-Wissenschaftsverlag, Mannheim.

Pfeifer, W. (1998), *An Introduction to the Interacting Boson Model of the Atomic Nucleus*, vdf Hochschulverlag an der ETH, Zürich.

Rotenberg, M., Bivins, R., Metropolis, M. and Wooten Jr, J. K. (1959), *The 3-j and 6-j Symbols*, Technology Press MIT, Cambridge, MA.

Talmi, I. (1993), *Simple Models of Complex Nuclei*, Harward academic publishers, Chur.

Index

9 783764 324186